IN THE FOOTSTEP
JOHN WESL
POWEL

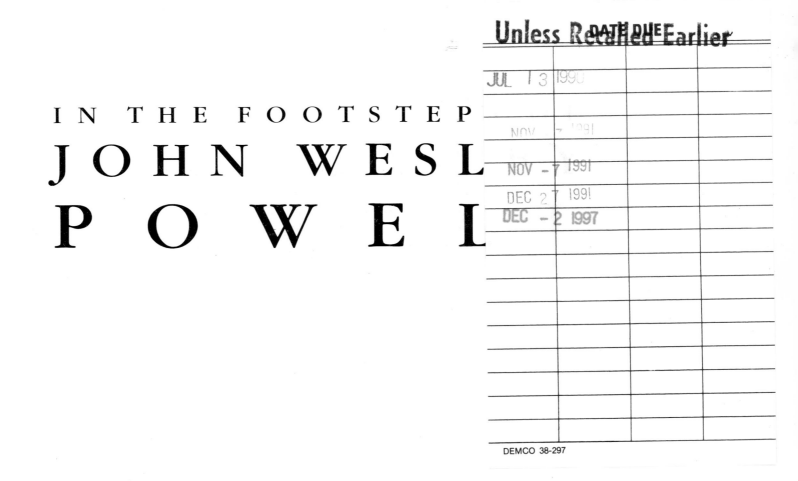

Gene Shoemaker (left) is plotting the position of Beaman's camera station 728 on Trin-Alcove Bend, 400 feet above the Green River. Hal Stephens photographs the scene. Often a panorama of pictures is recorded from high vantage points for later reference as is done in this instance.

IN THE FOOTSTEPS OF
JOHN WESLEY
POWELL

AN ALBUM OF COMPARATIVE PHOTOGRAPHS OF THE GREEN AND COLORADO RIVERS, 1871-72 AND 1968.

HAL G. STEPHENS AND EUGENE M. SHOEMAKER

Johnson Books: Boulder
The Powell Society: Denver

Text and design of the book © 1987 by The Powell Society, Ltd.

Printed in the United States of America by
Johnson Publishing Company
1880 South 57th Court
Boulder, Colorado 80301

Cover design by Molly Davis

ISBN: 1-55566-020-7 (cloth)
ISBN: 1-55566-025-8 (paper)
LCCCN: 87-82990

The Powell Society, Ltd., is a Colorado corporation, not for pecuniary profit, which has as its focus of interest the canyons of the Colorado River and particularly the work of John Wesley Powell. The society may be reached at 777 Vine Street, Denver, Colorado 80206. The trustees of the Powell Society are David Gaskill, Robert Gaskill, Felix Mutschler, George Ogura, Marvis Ogura, George Simmons, Henry W. Toll, Jr., and Lydia Toll.

Contents

Maps

Tables

Geologic Formations Mentioned in Text, Grouped
According to Segments of the 1968 Expedition

Foreword

The pioneering photographs of the second Powell expedition, taken by E.O. Beaman, James Fennemore, and then Jack Hillers in Grand Canyon, have not received adequate attention from historians of the West. To be sure, Powell did use both photographs and engravings taken from the photographs in the early accounts of his discoveries. However, several years later, he returned to the Canyon country with two of the most gifted artists of the nineteenth century: Thomas Moran and William Henry Holmes. And it is the great panoramic views of Moran and Holmes, instead of the photographs, that have become synonomous with the romance of John Wesley Powell and the canyons of the Colorado.

The authors have performed an important service to history by gathering the remaining photographic plates of the Powell expedition from old files and museum basements, bringing them together in chronological sequence, and explaining to us the near-heroic physical feats involved in lugging a ton or more of wet-plate materials, chemicals, and darkroom equipment through the depths of the Colorado canyons.

But this book is more than history; it honors the spirit of John Wesley Powell by putting those old glass plates into the service of modern geological inquiry. The authors set out to relocate the exact sites from which the photographs were originally taken, and then to shoot identical views. Some of the record has of course been obliterated by the works of man, specifically the great dams at Flaming Gorge and Glen Canyon. But in long stretches between the dams and reservoirs, the river still flows through landscapes largely undisturbed since the days of the Powell expedition. And in these sections of the river, the authors have provided us an unprecedented record of exactly what happens in a century of geological time.

The answer, in the unblinking black and white objectivity of the photograph, is . . . not very much. Here and there a boulder has fallen from a ledge, rock has spalled from a cliff face, some rock edges appear a little more rounded. The river tributaries, accumulating and carrying the erosional debris of thousands of square miles, tell us a somewhat more dynamic story. Here and there a side canyon, clean in the 1871 photograph, is now choked with boulders carried down by a sudden flash flood. At Warm Springs Rapid on the Yampa River, the photographs show us how a tributary, dumping its load in the river, has created a larger rapid.

Along the river itself, we see more change along banks and sandbars where the river relentlessly shifts, sorts, and reshapes its load of rocks and sand. But even in these remote reaches, the most obvious change is man-caused; the dense thickets of tamarisk lining the river banks were nowhere in sight when Powell's exploring parties passed through. Tamarisk, a Middle Eastern exotic planted for windbreaks in the Imperial Valley around 1900, has now spread throughout the entire Colorado River system.

On rare occasions, a lucky visitor will witness the changes that these pictures document. In October of 1980, I was with a group climbing out of Marble Canyon from President Harding Rapid up to Eminence Break. Suddenly the morning stillness was broken by a muffled explosion, and we looked down to see a huge cloud of dust rising from a rock slide breaking out of the red wall and cascading onto a talus slope at river's edge.

However, for our purposes, and on our time scale, the reassuring testimony of the century of time captured in this book is that the canyons of the Colorado are virtually eternal. What does it matter that in some time unfathomable to us the canyons will finally erode away? The task for us, in the meantime, is to keep the works of man out of the process, leaving time and the river to their geological destiny.

Bruce Babbitt
June 2, 1987

Acknowledgments

Many highly talented people at the Flagstaff Field Center of the U.S. Geological Survey (USGS) and elsewhere contributed substantially to the preparation of this album.

James W. Van Divier, Roger Carroll, Ramon Sabala, and Hugh Thomas (all USGS) retouched the Beaman and Hillers copy negatives and prints, skillfully removing spots, blemishes, and evidences of cracks that appeared in some of the historic plates. Their contributions to the enhancement of these photographs are much appreciated.

The film positives provided by the U.S. National Archives were copied onto 8″ x 10″ negative material by Karl Zeller. Ramona Boudreau also demonstrated superb photographic skill in matching and processing the Beaman and Hillers photographs. (Both Karl and Ramona are with the USGS.) Their work is gratefully appreciated.

Russell Wahmann (USGS) and Mary Young (privately employed) performed the exacting work of map preparation. The authors are indebted to them for their valuable contributions.

Technical reviews of the manuscript were provided by George C. Simmons (USGS, retired) and J. Phillip Schafer (USGS). The geologic information and historical background that they added to the manuscript were extremely helpful.

Arthur M. Phillips III, Curator of Botany for the Museum of Northern Arizona, Flagstaff, examined the vegetation seen in the photographs to verify or correct the field identifications. The authors gratefully acknowledge Dr. Phillips's contribution.

Special appreciation goes to Doris Weir, assisted by Susanne Bounds (both USGS), for the dedication and skill they brought to the final editing of the manuscript.

Our first draft reflected the geologic nomenclature of 1968. The authors are indebted to Marjorie E. Mac Lachlan (USGS, Denver) for updating it.

The many drafts were typed by Virginia M. Hall (USGS), whose accuracy and efficiency continue to impress us. Our heartfelt thanks go to her.

Above all, our thanks go the crew members who accompanied us on the expedition and lent valuable assistance.

First Segment: Patrick Shoemaker, camp assistant
Second Segment: E. C. Morris (USGS), geologist
                 Phil Hayes (USGS), geologist
                 Elmer Santos (USGS), geologist

	Carolyn S. Shoemaker, cook
	Patrick Shoemaker, camp assistant
	Elliot Morris, camp assistant
Third Segment:	Orson Anderson (UCLA), physicist
	Chester Anderson, boatman
	Carolyn S. Shoemaker, cook
	Patrick Shoemaker, camp assistant
	Linda Shoemaker
Fourth Segment:	Felix Mutschler, geologist
	Thor Karlstrom (USGS), geologist
	Karl Karlstrom, cook
	Patrick Shoemaker, camp assistant
	Richard Chidester, camp assistant
Fifth Segment:	George Simmons (USGS), geologist and boatman
	Dave Gaskill (USGS), geologist and boatman
	L. T. Silver, professor of geology, California Institute of Technology, Pasadena
	Bruce Julian, graduate student, California Institute of Technology
	Joan Anderman, cook
	Henry Toll, physician and boatman
	Tad Nichols, photographer
	Erling Jensen, Nichols's assistant
Sixth Segment:	George Simmons (USGS), geologist and boatman
	Dave Gaskill (USGS), geologist and boatman
	Maurice Brock (USGS), geologist
	George Anderman, geologist
	Joan Anderman, cook
	L. T. Silver
	Bruce Julian
	George Ogura, physician
	Henry Toll, physician and boatman

We also thank Tad Nichols of Tucson, Arizona, who met our river party at the lower end of Cataract Canyon and, using his power boat, towed us many miles to Hite, Utah.

Introduction
Background of Powell's Second Expedition

Major John Wesley Powell made two voyages of exploration down the Green and Colorado rivers. Although the first trip in 1869 was more famous and widely publicized, the second traverse in 1871-72 was better documented and produced more scientific results.

Powell's Smithsonian Report of 1875—his official report—included material from both expeditions, but observations were given in diary format as though made during the 1869 trip. Members of the 1871-72 expedition were ignored in the 1875 report. Frederick S. Dellenbaugh, who became the unofficial chief chronicler and historian of the second expedition, published two books and many articles concerning it, yet he once found it necessary to obtain a letter from Powell certifying that he was indeed a member of the party.

The first expedition left Green River Station, Wyoming, in May 1869 and arrived a little over 100 days later at the mouth of the Virgin River in present-day Nevada, about a day's journey by boat from the lower end of the Grand Canyon. Major Powell was the only scientist in the ten-man group. Most of the party were mountain men and trappers whom he had recruited in Colorado the previous year. They had expected to supplement their food stores by hunting, but much of the canyonlands through which they passed proved to be nearly barren of game. Thus they were forced to depend largely on what they carried with them, chiefly flour and bacon that Powell had acquired as army rations. A disaster in Canyon of Lodore early in the trip resulted in the loss of one of their four boats—the one that contained much of their food. Consequently, the remainder of the voyage was extremely hurried; the men who remained with Powell probably were suffering from scurvy and were near starvation when they reached the mouth of the Virgin River.

Powell's observations and notes were far too sparse to meet his objective of plotting the course of the river and the topography of the adjacent country. The great haste required to complete this voyage of survival precluded his spending sufficient time to acquire scientific and topographic data. He was also unable to have photographs taken on the first expedition because he lacked the necessary funds. He had financed the trip out of his own pocket and with the meager funds appropriated by the Illinois Legislature.

Immediately following the first trip, Powell began seeking additional money and planning a second expedition over the same route. His dramatic journey in 1869 had catapulted him into national fame, and it was largely responsible for his obtaining funds from Congress to continue his scientific explorations. He spent much of 1870 exploring routes by which supplies could be carried to various points along the rivers

1

and cached for later use during a much less hasty voyage; he needed adequate time for a systematic topographic and geologic survey of the Colorado River and its tributaries.

Jacob Hamlin, a Mormon missionary familiar with the territory between the new settlement of Kanab, Utah, and the mouth of the Paria River, helped Powell in his exploration of the tributaries of the Colorado River during the fall of 1870. Hamlin's knowledge of the Shivwits Indians and their language enabled Powell to learn the details of the deaths of William Dunn and the Howland brothers. They had left the party at Separation Rapids on the Colorado River in 1869 and had been killed by the Shivwits.

In May 1871 Powell began his second expedition with an entirely new crew of ten men. The departure point was the same as for the 1869 trip—Green River Station, Wyoming. The crew included Professor Almon H. Thompson, Powell's brother-in-law, who was in charge of topographic work, assisted by Francis M. Bishop, a Union Army veteran, and Stephen V. Jones; E. O. Beaman, of New York, who was the photographer, assisted by Walter C. Powell, Major Powell's first cousin; and John F. Steward, a geologist. The boatmen were seventeen-year-old Frederick S. Dellenbaugh, who was also the expedition's artist; Andrew J. Hattan, who had known J. W. Powell in the army and who doubled as cook; and John K. Hillers, who became the expedition's photographer in 1872. The tenth member, Frank Richardson, left the party at Browns Park, near the beginning of the journey. All of the expedition members, except Beaman, were also amateur scientists. The three wooden boats specially built for the voyage are shown in many of the photographs in this album. They were named the *Emma Dean*, after Major Powell's wife; the *Nellie Powell*, after J. W. Powell's sister; and the *Cañonita*.

Photography in the 1870s

The wet-collodion process of photography used on the expedition was first devised in the 1850s and refined during the 1860s. This process was developed primarily for astronomical photography, but it soon became popular for general photographic purposes because it permitted the printing of an unlimited number of copies from a single negative plate.

Collodion, containing a mixture of bromides and iodides of cadmium, potassium, and ammonium, was used to coat a clean glass plate. When dry, this plate was dipped in a silver nitrate solution saturated with silver iodide. The bromides and iodides in the collodion were converted to silver salts and the plate was ready to be exposed while wet—within ten minutes of preparation. The latent image was immediately developed in a solution of iron sulfate and acetic acid. Alcohol was added to allow penetration of the developer into the collodion. The images thus developed could be intensified in an acid solution of silver nitrate, either immediately or after fixing. (A dry collodion plate could be used if all excess silver nitrate was washed off before drying; however, the dry emulsion was many times less sensitive than the "wet plate.") When executed skillfully, the wet-plate process produced photographs of astonishingly good quality and fine grain.

This wet-plate process required that a darkroom tent and all necessary processing chemicals be transported to the camera stations, together with the camera and glass plates. Diaries of the expedition members contain many references to the onerous task of carrying all the equipment to lofty camera stations. Even on the river, the task of transporting and keeping dry all this material must have been formidable—there was nearly a ton of it!

The Second Powell Expedition and Its Photographers

This expedition was planned to extend over a period of about a year and a half, and the Major adhered closely to this schedule. About six months were required to traverse from Green River Station, Wyoming, to the mouth of the Paria River at the foot of Glen Canyon in Arizona, the site of Lees Ferry. A. H. Thompson was responsible for most of the topographic work done during these six months; Powell spent much of this period on horseback, traveling between the Green River and Salt Lake City and exploring the canyonlands west of the river.

During the winter of 1871-72, two boats were left at the mouth of the Paria, and the party established a camp just outside Kanab, Utah. During the winter, they surveyed a baseline and established a triangulation network for topographic control along the west and north sides of the Colorado River. Powell also spent part of this time studying the local Indian tribes.

Beaman, the photographer, became disenchanted with the expedition during the winter and, following a disagreement with Major Powell, elected to leave the party in January 1872. Most of the photographic equipment already belonged to Powell, and he bought out Beaman's share of the photographs that had been taken thus far. Beaman eventually crossed the Colorado River at Lees Ferry and traveled eastward through the Navajo and Hopi country.

G. C. Simmons contributed an account of Beaman's photography in the Grand Canyon and the side canyon of Kanab Creek. After leaving Powell, Beaman went to Salt Lake City and purchased photographic equipment. He returned to Kanab and hiked down Kanab Creek canyon to the Colorado River. He then hiked up the Colorado with a Sam Rudd to Buckskin (now Deer Creek) Falls and up over the cliffs into Surprise Valley. The photographs Beaman took on his journey predate by a few days those taken by Powell's expedition; thus Beaman was the second to photograph the Grand Canyon (Timothy O'Sullivan having been the first). Beaman returned to Buckskin Falls, constructed a small log raft, and floated alone back to Kanab Creek canyon, also predating Powell's expedition on this portion of the Colorado. Beaman then hiked back to Lees Ferry, but was unable to cross into Navajo country because of high water, so he went to the Buckskin Mountains on the Kaibab Plateau and obtained pictures of the Colorado from the North Rim. The account of his journeys was serialized in the now-extinct *Appleton's Journal.*

A new photographer, James Fennemore, was recruited in Salt Lake City to replace Beaman. He remained with the Powell party for only a few months, as ill health forced him to leave before the river trip was resumed in the latter half of August 1872. One of his major contributions to the expedition, however, was his training of Hillers, who then assumed the photographic responsibilities. In April 1872, while the rest of the party were at Kanab or off on their own individual adventures, Hillers accompanied Fennemore into the Grand Canyon to photograph Lava Falls. Then, in July 1872, Fennemore took a number of photographs in Glen Canyon—now submerged beneath Lake Powell.

The brief apprenticeship with Fennemore started Hillers on a promising photographic career. Hillers took the remainder of the pictures on the river trip—all of those in Marble and Grand canyons, and several in Kanab Creek canyon. Hillers remained with the Powell survey as its photographer and later became Chief Photographer for the U.S. Geological Survey. He retired from the Survey shortly after the turn of the century.

Importance of the Powell Photographs

The pictures acquired on the second Powell expedition were the basis for illustrations in Powell's Smithsonian Report. Because the halftone process had not yet come into use, Powell used reproductions of line drawings made by artists from the photographs. These included scenes at various points in the upper canyons and in Grand Canyon and several outstanding portraits of Indians made by Hillers. Some liberties were taken by the artists who prepared the line drawings. In composing the pictures of the canyons, they often exaggerated the height and steepness of canyon walls.

The pictures also played an essential role in lecture tours that Powell made during the following years. These tours, one of them with Hillers as projectionist, provided Powell and Hillers with considerable income. In addition, Powell sold reproduction rights for a large number of stereophotographs, which were distributed throughout the country for home use in the then-popular stereopticon viewer.

Approximately 225 pictures or stereopairs of pictures suitable for reproduction were made by Beaman, Fennemore, and Hillers. About two-thirds of these were taken by Beaman, a few by Fennemore, and the remainder by Hillers. Most of the pictures are stereopairs taken on plates about 5″ x 8″; each half of the pair measures 4″ x 5″. A smaller number of pictures were taken on the large 11″ x 14″ plates. Most of the views are of the canyonlands country along the rivers or of the rivers taken from the canyon rims. One or more members of the Powell expedition are included in twenty-one of the photographs, and one portrays the entire group in the boats at the start of the second expedition.

Most of the photographs in the collection are excellent, and some are superb. The best of them are as good as any obtained by W. H. Jackson, famous for his western scenes. Both Beaman and Hillers deserve to be recognized among the giants of early western photography.

Less than 100 plates from the original set still exist. Some are in good condition, but many are not. Most have pinholes and scratches; some are chipped and broken. Nearly all the pictures have the spots or blemishes that were inherent in the old wet-plate process. Varnish was often used to preserve the surface emulsion, and brush marks are conspicuous on many of the plates. The skyline in many of the photographs was altered when the sky was opaqued to remove obvious defects in the plate emulsion.

During the 1930s, many of Hillers's plates and some of Beaman's were destroyed to conserve critical storage space at facilities of the U.S. Geological Survey. Reference prints of excellent quality, however, had been made by the Survey from these plates. The prints were photocopied at the Survey's Flagstaff, Arizona, Field Center, and some have been selected for this album.

The remaining original photographic plates (exclusive of pictures of Indians) compose the Hillers Collection, which now resides in the National Archives. Copy negatives of these plates filed in the U.S. Geological Survey archives in Denver, Colorado, along with the prints from the destroyed plates, provided the source of reference prints used on our river trip in 1968. The identifications of the pictures have become scrambled over the years. The only direct information that we have on the pictures consists of captions and sparse marginal notes in the U.S. Geological Survey's reference album of the Hillers Collection. In some cases, even these captions proved to be in error.

The photographs of Powell's second expedition down the Green and Colorado rivers have never been systematically published. The most extensive published use of the photographs was made by Frederick S. Dellenbaugh in his two books, *The Romance of the Colorado River* and *A Canyon Voyage,* both of which appeared more than thirty years after the time of the expedition. The pictures were reduced in size in these books and hardly do justice to the quality of the original photographs. Reprints of the books by the offset method in recent years have further degraded the quality of the pictures. The public thus has had very little opportunity to examine these historic photographs in anything near their original state.

A few of the plates from the 1871-72 expedition, especially those by Hillers, have been used in various publications of the U.S. Geological Survey. Most of the Beaman and Hillers pictures, however, have never been seen in any form by the public, except possibly as stereopticon pictures circulated more than three generations ago.

The most reliable source materials left today to aid in identification of the Beaman, Fennemore, and Hillers pictures are the diaries and journals kept by the various members of the party. A surprising amount of factual information is contained in some of these accounts. Journals were kept by Major Powell, Thompson, Bishop, Jones, Steward, W. C. (Clem) Powell, Hillers, and Dellenbaugh. Of these, the journals of Thompson, Bishop, Jones, Steward, and W. C. Powell have been published by the Utah Historical Society. Dellenbaugh published his own account in his two books, particularly in *A Canyon Voyage,* which he intended to be the principal narrative record of the expedition. The book, however, is not in diary form and is not very useful as an account of day-to-day activities of the trip. His unpublished diary, upon which *A Canyon Voyage* was based, contains a great deal more of specific information.

Powell's journal was deposited in the Bureau of Ethnology of the Smithsonian Institution and may some day be published. Hillers's diary was published by the University of Utah Press (Fowler, ed., 1972). It does provide a daily account of the expedition and is quite detailed in places.

The 1968 Expedition: Purpose, Organization, and Photographic Techniques

The U.S. Geological Survey, in conjunction with the Smithsonian Institution and the National Geographic Society, planned to commemorate in 1969 the centennial of Powell's first river voyage. As part of the Survey's participation in the centennial, E. M. Shoemaker proposed to the Survey to retraverse the entire length of Powell's second expedition down the Green and Colorado rivers. The major objectives were to relocate as many of the Beaman and Hillers camera stations as possible, identify positively the pictures taken from the stations in 1871-72, and rephotograph the scenes from the original camera stations. The plan was for H. G. Stephens to accompany Shoemaker over the traverse and to be responsible for the photography.

The proposal was approved by the Survey and the logistics of the three-month 1968 expedition were planned forthwith. The Green and Colorado rivers were divided into six geographic segments, and about two weeks of field work were allowed for each segment. Fifteen-foot neoprene inflatable rafts, sometimes rigged with rowing frames and oars, and sometimes propelled with canoe paddles, were used—two on the upper parts of the Green and Colorado and five in Marble and Grand canyons. An eighteen-foot power boat was used to locate the positions of camera stations submerged by Flaming Gorge Reservoir.

The crews of our river party changed from segment to segment. They included members of the U.S. Geological Survey from Denver and Flagstaff and veteran river runners from Denver, Pasadena, and elsewhere (their names are listed in the acknowledgments). Several of the participating geologists prepared detailed geologic logs of various portions of the river, which have since been published by the Powell Society, Ltd., of Denver (see Selected Reading).

Field work on the river centered around the problem of locating each Beaman or Hillers camera station. A file had been prepared in advance, containing data folders devoted to each day of Powell's expedition.

For a given day, the folder contained copies of all the journal entries by Powell expedition members, prints from the Hillers Collection, and maps showing probable locations of camera stations and Powell's campsites occupied on that date.

Our detailed knowledge of the canyon geology was a major factor in locating the old camera stations. Once we recognized the general geologic features in a Beaman or Hillers photograph, we landed and determined the exact camera site on the river bank by study of the reference picture and analysis of parallax between features in the near and far fields. When we were able to identify positively the foreground features in the photographs, we could usually find the camera station within a few inches of the exact spot. While Shoemaker prepared descriptive notes of the site, Stephens placed his camera on a tripod as near to the original camera position as possible and took a duplicate picture.

At a few places on the river banks, Stephens could not occupy the exact station because it had been changed by the river, a tributary stream, or a rock fall. In the Canyon of Lodore, for example, a ten-ton boulder from the high canyon wall now reposes on the station where Beaman had set up his camera to obtain photograph 504. The 1968 photographs, in such cases, were taken from an offset position, the extent of which is noted in their captions. At other places, we encountered heavy growths of tamarisk (introduced into the region some time around the turn of the century), box elder, or cactus that obscured the view from a formerly barren camera site. At one station on Kanab Creek, that cactus was not to be argued with! It was impossible, of course, to reoccupy sites now flooded by Flaming Gorge Reservoir and Lake Powell. We were able to duplicate roughly some of the gorge scenes, but the lake has drowned the Glen Canyon of Powell's day so completely that we omitted the 162 river miles between Hite, Utah, and Lees Ferry, Arizona.

It was unfortunate that the tight schedule of our trip did not permit us to photograph all the scenes at the ideal time of day. Especially in narrow side canyons, contrasts in light were at times so strong that details were lost in shadow. Conversely, to photograph some foreground details properly, we had to sacrifice the clarity of distant features. In this respect, Powell's photographers were sometimes more fortunate than we were. In almost all cases, however, where one picture has lost detail, its mate has not, and the reader can visualize the scene by studying the two together.

Stephens used a Graflex XL camera with a 2¼″ x 3¼″ format to rephotograph the scenes. His lens was an ultra-wide-angle lens, a Schneider Super Angulon of 47-mm focal length. For all exposures, he used Eastman Kodak Plus-X roll film, 120 size, rated at 160 ASA, and a medium-yellow filter. (Use of trade names is for descriptive purposes only and does not imply endorsement by the U.S. Geological Survey.)

Stephens's use of only the wide-angle camera came about by accident. At the beginning of our expedition, we did not realize that Beaman and Hillers had used a camera with extremely wide-angle coverage, and we planned to do our work with a standard-lens camera. We did have a back-up camera with us, though—luckily a wide-angle one. When Stephens, who was carrying the standard camera on his back, fell on it while climbing steep slopes in Canyon of Lodore, the wide-angle camera was all we had. It proved, after all, to be the one best suited to matching the 1871-72 pictures.

Preparation of the Album

Altogether, we identified about 150 camera stations of the Powell expedition. Of the resulting sets of photographs, we selected 110 for this album. We cropped and scaled our 1968 pictures to obtain a reasonably good match of the original pictures. Reference numbers assigned by the U.S. Geological Survey many years after the expedition are used by the National Archives to identify the Beaman-Hillers photographs. These numbers are retained in this album.

Restoration of the old photographs taken by Beaman and Hillers was a complex and difficult process. We were fortunate that it could be done in the Survey's photographic laboratory at Flagstaff, whose personnel had been processing the images obtained from lunar and planetary exploration and were familiar with techniques that would both enhance the photographs and remove artifacts. The laboratory produced 8″ x 10″ copy negatives from the contact positives made by laboratories at the National Archives. These negatives were retouched, and 11″ x 14″ positive prints were made from them in order to remove the undesirable effects of both opaque and clear blemishes. The 11″ x 14″ prints became the final product for publication. We used 8″ x 10″ copies of these prints for review and editing.

Introductory sections precede the photographs reproduced for each segment of the 1968 expedition. They are designed to provide continuity to the series of paired photographs and to acquaint the prospective river traveler with the character and mood of each canyon traversed; geologic comments here are of a general nature. (We have provided at the back of this book a glossary of terms, largely geologic, that may be unfamiliar to the general reader.) Captions under the pairs of pictures describe in greater detail the rocks viewed in each scene. (The river runners' guides published by the Powell Society, Ltd., listed in Selected Reading, provide a continuous account of the geologic formations and structures along the rivers together with historical background.) The dates given with the Powell pictures were obtained or inferred from the journals of the expedition members; therefore, some dates given are more precise than others.

Formally named geologic units that we describe are tabulated at the beginning of each segment of our 1968 trip so that the reader may know the proper position of each unit in the stratigraphic column. The units are listed in conventional order (youngest at top, oldest at bottom). The tables include *only* those units mentioned, and some that may be present in the area are therefore not listed. (For example, in Table I we show the Dakota Sandstone on the Curtis Formation, but in Table II the Morrison Formation rests on the Curtis Member of the Stump Formation. A complete geologic section would include the Morrison Formation and the overlying Dakota Sandstone.) Nomenclature of the units has changed considerably since the 1870s as the geology of the region has become better understood. We use the names presently accepted by the U.S. Geological Survey, but we have included Powell's original nomenclature. A complete geologic time scale is given at the back of the book.

We feel certain that some readers will be interested in studying further the topography of individual areas. Therefore, with each of the 1968 pictures we give the name and size of the pertinent topographic quadrangle map. The scale of the quadrangles covering 7½ minutes of latitude and longitude is 1:24,000 (1 inch = 2,000 feet); that of the 15-minute quadrangles is 1:62,500 (1 inch = about 1 mile). All of these maps in the four states may be ordered from the Branch of Distribution, U.S. Geological Survey, Box 26286, Federal Center, Denver, CO 80225. Also available from this source are four special-area topographic maps that cover parts of the river traverse: Canyonlands National Park and vicinity, Utah; Dinosaur National Monument, Utah-Colorado; Glen Canyon Recreational Area, Utah-Arizona; and Grand Canyon National Monument, Arizona. (Grand Canyon National Monument is now part of Grand Canyon National Park.) Our usage and spelling of place names are those of the topographic maps.

A trip through the canyons of the Green and Colorado rivers provides the traveler with views of some of the most extraordinary exposures of geologic sections seen anywhere in the world. Beginning in rocks of Tertiary age, less than 60 million years old, at Green River, Wyoming, one travels nearly a vertical mile down through the geologic column to Early Proterozoic (Precambrian) exposures in the Inner Gorge of the Grand Canyon that are more than 1.5 billion years old. Through the photographs and notes in this album, we hope to convey the character of the rugged canyon country, the awesome views, and the fearsome but exhilarating rapids that confronted John Wesley Powell and his courageous party during their history-making explorations of the Green and Colorado rivers. Tens of thousands of people have run the Colorado River through the Grand Canyon since the time of Powell, but it remains one of the most adventurous and exciting experiences of our time.

Index map showing Green and Colorado rivers traversed by the second Powell expedition of 1871-1872 and 1968 expedition segments

North Half

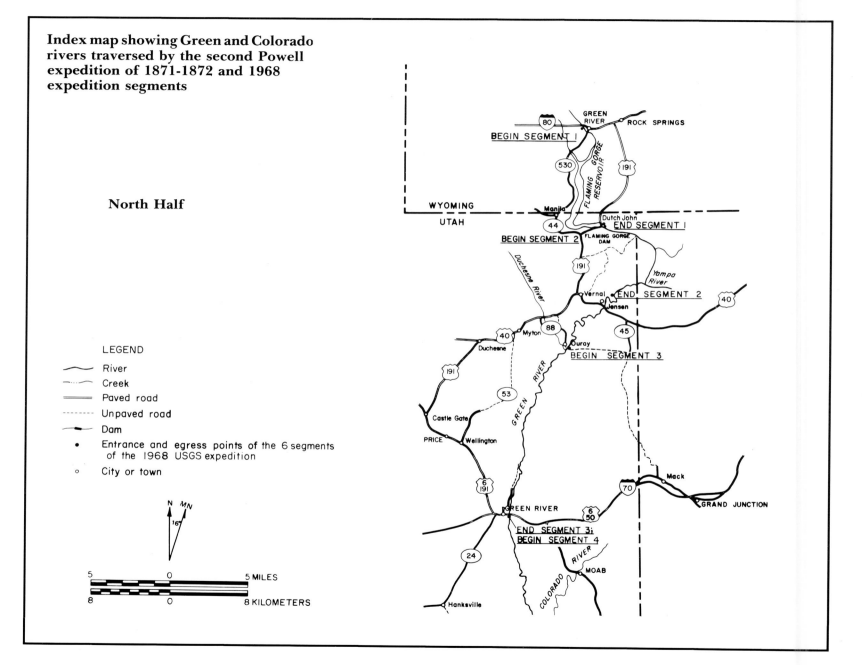

LEGEND

〰 River

·—·— Creek

═══ Paved road

- - - - Unpaved road

▬■ Dam

• Entrance and egress points of the 6 segments of the 1968 USGS expedition

○ City or town

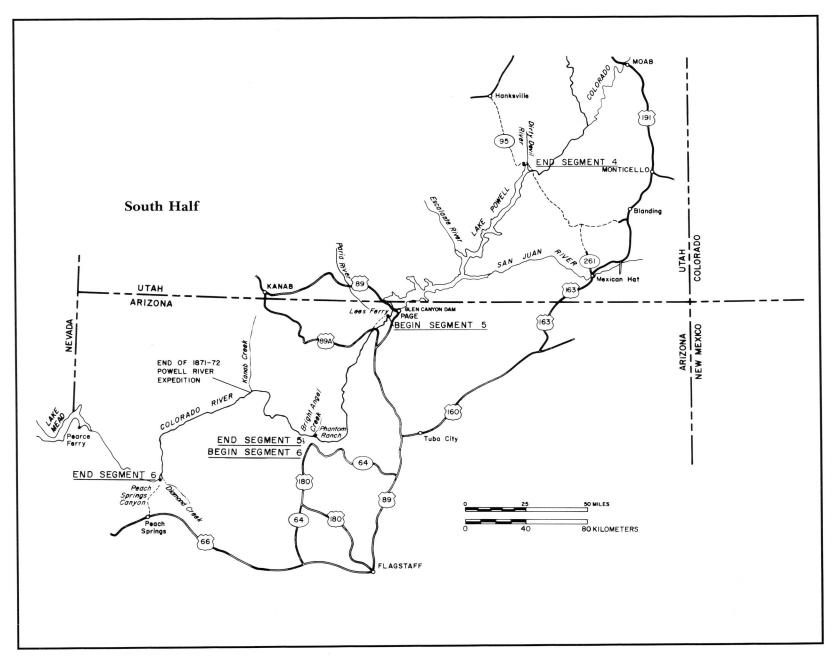

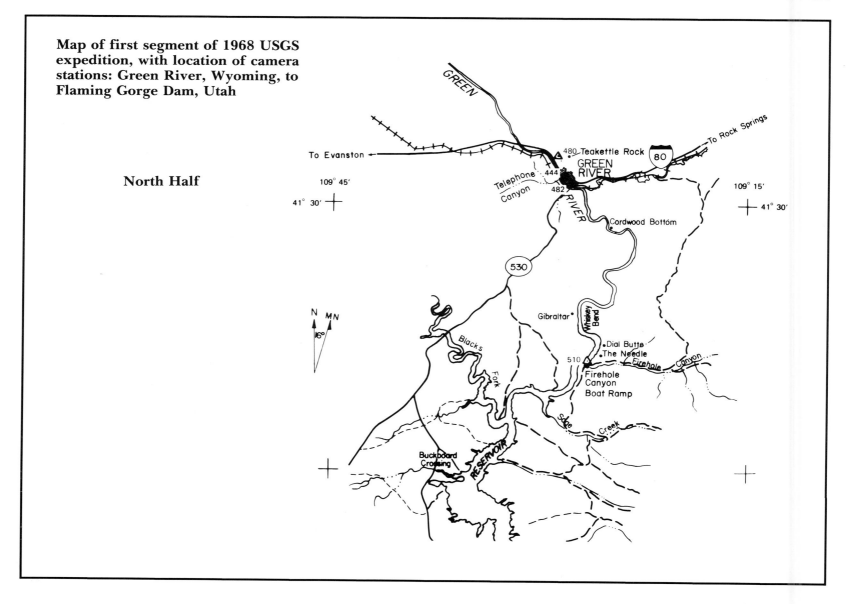

Map of first segment of 1968 USGS expedition, with location of camera stations: Green River, Wyoming, to Flaming Gorge Dam, Utah

North Half

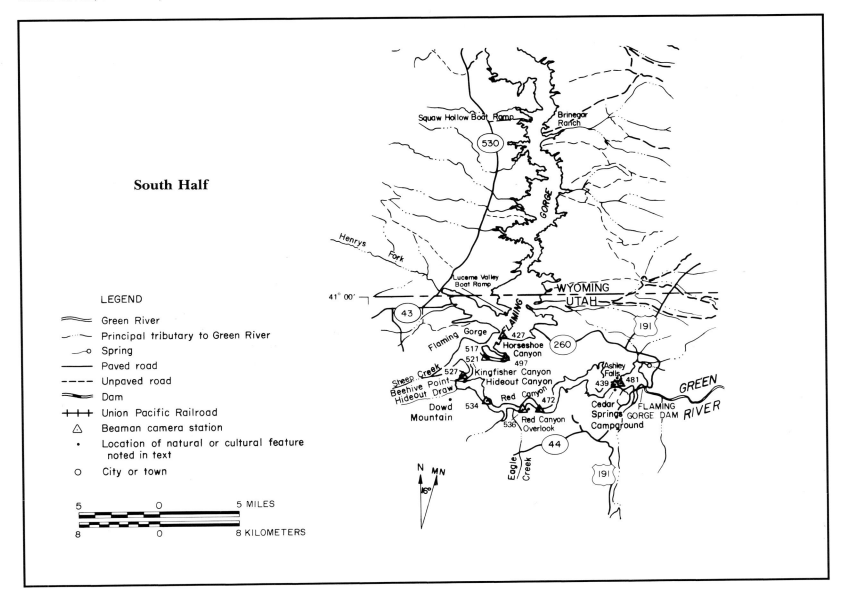

South Half

LEGEND

〰 Green River
·-·-·-·- Principal tributary to Green River
⌒о Spring
——— Paved road
----- Unpaved road
〰 Dam
+++ Union Pacific Railroad
△ Beaman camera station
· Location of natural or cultural feature
   noted in text
O City or town

5   0   5 MILES
8   0   8 KILOMETERS

Table I: First Segment, 1968 Expedition

Geologic formations mentioned in captions of photographs taken between Green River, Wyoming, and Flaming Gorge Dam, Utah.

Geologic Era	Geologic Period	Formation or Rock Type	Remarks (Numbers are those of Beaman Photographs)
CENOZOIC	QUATERNARY	Pleistocene alluvium and talus	Alluvium cut by gully in foreground of 480
	TERTIARY	Green River Formation	Dominant rock formation in vicinity of Green River, Wyoming.
		Laney Member	In 480 and others in northern part of Flaming Gorge Reservoir as far south as Horseshoe Canyon.
		Wasatch Formation	Near Flaming Gorge entrance (427).
MESOZOIC	CRETACEOUS	Dakota Sandstone	Near Flaming Gorge entrance (427); also in 517.
	JURASSIC	Curtis Formation Entrada Sandstone Carmel Formation	
	JURASSIC AND TRIASSIC	Glen Canyon Sandstone (has been called Navajo Sandstone or Nugget Sandstone by others)	Exposed in Flaming Gorge and shown in 427.
	TRIASSIC	Chinle Formation Gartra Grit Member Moenkopi Formation	
PALEOZOIC	PENNSYLVANIAN	Weber Sandstone	Walls of Horseshoe Canyon (517), downstream to 534.
MIDDLE PROTEROZOIC		Uinta Mountain Group	Forms Flaming Gorge walls from 536 on; first seen in 527.

Chapter One

Green River, Wyoming, to Flaming Gorge Dam, Utah

The 1968 U.S. Geological Survey expedition set out on July 5. Our first task was to examine and photograph the vicinity of John Wesley Powell's departure point for his 1869 and 1871-72 river voyages.

The Green River Station of 1871 has grown in the past century from a village of less than a hundred inhabitants into a small city of about 4,200 and is now known simply as Green River. The town is on Interstate Highway 80 near the southwest corner of Wyoming and lies at an altitude of 6,000 feet. It may be reached from the west on Interstate 80 out of Salt Lake City, Utah, or from the east on the same highway passing through Cheyenne, Laramie, and Rawlins, all in Wyoming. Access from the south is from Vernal, Utah, on U.S. Highway 191, connecting with Utah State Highway 44. This road skirts the southwest shores of Flaming Gorge Reservoir, becomes Wyoming State Highway 530 farther north, and continues along the west side of the reservoir to Green River. The water of Flaming Gorge Reservoir floods Green River Canyon to within eight miles of the town.

The surrounding area is dominated by the browns and grays of the Green River Formation, and ledges and buttes are prominent. Powell gave the name Tower Sandstone to the unit forming these resistant features, but the name has since been abandoned in favor of the Laney Member of the Green River Formation. High-desert scrub and bunch grass dot the hillsides. From the town of Green River to the backed-up waters of Flaming Gorge Reservoir, the canyon country is drab in color and broadly sloping, except where rust-brown cliffs, ledges, and buttes are most common. The upper, northern part of the reservoir has wide expanses of bays and inlets bordered by gently sloping canyon walls.

Farther south, about three miles into Utah, the landscape changes dramatically. Fiery red sandstones and shales of the 1,000-foot cliffs at the entrance to Flaming Gorge proper mark the beginning of the narrow, sinuous portion of the lake as it twists through Horseshoe and Red canyons. These canyons are steep walled and heavily timbered, providing a stark contrast to the gentle, nearly barren slopes uplake. Farther down, the Green River of 1871 traversed the flanks of the Uinta Mountains and near Red Canyon Overlook had cut into the ancient dark-red quartzite of the Uinta Mountain Group. The present lake, now narrowed, follows this former tortuous course to Flaming Gorge Dam, the end of the first segment of our river voyage.

Green River
Camera Station 480

May 22, 1871

This pair of photographs and the next pair were taken just upriver from the departure point of the second Powell expedition. The camera station is on the west edge of a gully cut into alluvium of late Quaternary age. We are about 600 feet from its mouth on the Green River, looking northeast. The Green River Formation is exposed in the foreground on the floor of the gully and on the east bank. The gully seems to have widened and deepened during the past century, and its walls are steeper. Our difficulty in determining the precise location of the station (due to a large fill along Interstate 80 blocking the view) prevented an exact assessment of the amount of change.

We found the old eroded bridge abutment shown on the east bank of the gully in the foreground of the Stephens picture but saw no trace of a matching abutment on the west side nor any clear trace of a

July 6, 1968

Telephone Canyon in the background is cut in the Laney Member of the Green River Formation, which is capped by a resistant sandstone layer that forms promontories and cliffs. The 1871 gravel bar on the far side of the river just beyond the boats has extended to the full width of Stephens's photograph, and a luxuriant willow thicket on the bar now obscures much of the distant slopes.

Green River 7½' quadrangle, Wyoming

19

Green River
Camera Station 482

Probably May 22, 1871

The present-day setting of Beaman's camera station is just south of the Union Pacific Railroad switchyards and east of Wyoming State Highway 530. We could not locate the exact station because it has been covered by ten feet or more of cinders unloaded from the old coal-burning locomotives. The horizontal position of our station is probably within about ten feet of Beaman's, and we are looking west. All remains of the ruined buildings, including a stage station and repair shop, shown in his photograph have disappeared. We could discover no trace of adobe or other building materials that might date from the 1870s or earlier, but the foundations of the ruins might be found by excavation. A bluff on which the buildings once stood is easily visible from the present camera site. The modern paved highway, Wyoming State 530, extends from north to south down the distant canyon in the center of the picture. Part of the present-day town of Green River has been built on the southwest bank of the

20

July 6, 1968

river, on a terrace in front of our camera station below the skyline cliffs. A water tank now sits on the bench just above the chimney of the ruined building to the right of center in Beaman's photograph.

The bedrock in view is all Green River Formation. The skyline is formed by sandstone of the Laney Member of the Green River, and the slopes below are also Laney. The prominent butte in the right center of the pictures is formed by an unusually thick sandstone lens, one of several that occur at irregular intervals throughout the Laney that probably mark the sites of old stream channels. Shale beds in the Laney beneath these sandstone lenses are commonly warped, probably by compaction. (Shale is generally weaker than sandstone and is more likely to be compressed under the weight of overlying beds.)

Green River 7½' quadrangle, Wyoming

21

Firehole Canyon
Camera Station 510

May 24, 1871, about 9:00 A.M.

The original camera station is now under eleven feet of water, about fifty feet from the shore of Flaming Gorge Reservoir and 200 feet from Firehole Canyon Boat Ramp. The 1968 picture was taken from a boat, and the view is northeast. The two prominent spires were named by the Powell party: the Needle on the right and Dial Butte on the left. (Unfortunately, on the USGS topographic maps these spires have become North Chimney Rock and South Chimney Rock, respectively, indistinguishable by name from countless other "Chimney Rocks.") The intervening butte was referred to by Thompson as the "Boston Loaf Butte" and by Jones as "Boston Brown." All three features are capped by sandstone of the Laney Member, and their slopes are composed of a thick section of the Laney.

The Powell party spent considerable time studying the geologic section at these buttes. Thompson and Bishop measured their heights while Powell, Steward, and Jones correlated beds exposed here with those measured at Green River Station. Steward, writing

in his diary at the end of the day, became confused about the side of the
river on which Dial Butte is located
and referred to it as being on the opposite side of the river from the Needle.
His confusion is curious, as it was he
who suggested its name. This view of
both Dial Butte and the Needle from
some distance downstream may have
led him to think that the river passed
between them.

July 9, 1968

Firehole Basin 15' quadrangle, Wyoming

**Flaming Gorge,
Head of
Horseshoe Canyon**
Camera Station 427

May 28, 1871

Six Beaman photographs, numbered 478, 512, 459, 427, 496, and 908, provide a succession of views of a quarter-mile stretch of the west bank of the Green River at the entrance to Flaming Gorge (between the gorge and what is now the reservoir). All are morning pictures, probably taken in succession. Beaman photograph 427 is representative of all six. The stations are now about 140 feet under water.

Flaming Gorge was named by Powell for the brilliant red Triassic rocks that stand in vivid contrast to the drab greenish-gray shales of the Green River Formation through which his party had passed for fifty-odd miles. The gorge was formed where the Green River cut into the resistant rocks that are turned on end on the flank of the great Uinta Mountains arch. Steeply dipping beds extend for nearly four miles above the entrance of the gorge; an enormous thickness of rocks is thus exposed. Powell's camp no. 6 was at the upriver end (the stratigraphic top) of this exposed section, close to the contact of the Green River and Wasatch Formations. The Wasatch and all of the Mesozoic formations listed in Table I are turned on end along the river, but Beaman's photographs concentrate on the more spectacular Upper Jurassic to Lower Triassic rocks. (However, in the distance near the right border of

July 10, 1968

both pictures is a hogback of the younger Laney Member of the Green River Formation, partly obscured by foliage in Beaman's picture.) The Curtis Formation forms the upturned ledge on which we can see the power lines in the 1968 picture; here beds dip nearly 80° north. The next ledge to the left is formed by the much thinner Entrada Sandstone. The wide shaly saddle is Carmel Formation. To the left of it, the "Great Sandstone Ledge" of the Glen Canyon Sandstone forms a nearly vertical cliff about 1,000 feet high. Beneath the 36 cliff are brilliant red Triassic rocks. Topmost is the Chinle Formation; its basal unit, the Gartra Grit Member, forms a narrow sandstone ledge. Beneath it are the lighter red slopes of the Moenkopi Formation. At the extreme left in Beaman's picture, on the near side of the river, is a steep slope capped by cliffs; this slope is on a continuation of the Glen Canyon Sandstone. Note the bright streak in Beaman's photograph that begins about halfway up the cliff of Glen Canyon Sandstone that forms the highest skyline point; the streak extends down onto the talus below the cliff. This streak is a rock fall that must have been active at the time. In the last 100 years, the fall has extended upward to the top of the cliff, and a great slab of Glen Canyon Sandstone has come down from near the skyline.

Flaming Gorge 7½' quadrangle, Utah

Late May 1871

Horseshoe Canyon
Camera Station 497

Since Beaman's camera station is now about 150 feet under the water of Flaming Gorge Reservoir, we could not obtain a true replica of the scene. The station was on the inside of a bend of Horseshoe Canyon, on the west bank of the Green River directly opposite the ravine that is the subject of this picture. The entire wall of the ravine and all rocks in view are white, crossbedded sandstone of the Weber Sandstone. About a thousand feet of sandstone is exposed and in view of the camera.

We could detect no obvious changes in the canyon walls shown. Even the trees above the high-water line appear to be relatively unchanged. However, a rock slide to the right of the scene was active in 1871 and has been enlarged since; this slide is pictured in Beaman photograph 521.

July 10, 1968

Flaming Gorge 7½' quadrangle, Wyoming

Late May 1871

Horseshoe Canyon
Camera Station 517

This photograph and the one that follows, 521, probably were shot from the same camera station, now under about sixty feet of water. The station was on the south bank of the Green, on the head of a talus slope, slightly more than a mile downstream from the bend in Horseshoe Canyon. Photograph 517 provides an excellent downstream view of the "chimney rocks" about 250 feet to the west. Here the massive to cross-bedded sandstone of the Weber Sandstone is exposed on each side of the river. In the distance, just above the river and framed by the canyon walls, is the Glen Canyon Sandstone capping the flaming red scarp of the Chinle Formation (which includes its basal Gartra Grit Member) and the upper part of the Moenkopi Formation. A still more distant ridge in the extreme left of the gap between the canyon walls is capped by the much younger Dakota Sandstone, which just peaks above the nearer ridge of Glen Canyon. We could detect no changes on the chimneys in the left foreground. Even a little boulder perched on the right chimney has not moved in the last hundred years.

July 10, 1968

Flaming Gorge 7½' quadrangle, Wyoming

Late May 1871

Horseshoe Canyon
Camera Station 521

The location of this camera station is virtually the same as that of 517, the preceding one. Here we are looking upstream. The canyon walls shown, here more than a thousand feet high, are also Weber Sandstone. In the distance is the same ravine shown in Beaman photograph 497. In this picture we see an active rock slide in the upper two-thirds of the course of the ravine. It has been extended in modern times, particularly near the upper left edge. The man seated on a rock in the lower right of the Beaman photograph is probably Clem Powell.

July 10, 1968

Flaming Gorge 7½' quadrangle, Wyoming

May 31, 1871, 2:00 P.M.

Kingfisher Canyon
Camera Station 527

Beaman's camera station—now under 140 feet of water—was on the west bank of the Green below the mouth of Sheep Creek (called Kingfisher by the Powell party). The date and time of his picture were noted in Steward's journal. The view is southwest, across the river toward Beehive Point. Jointed, massive or crossbedded sandstone of the Weber Sandstone forms the wall of the canyon on the right; Stephens's photograph shows these beds arching up strongly at Beehive Point. Hideout Draw and Dowd Mountain are in the distance. What we can see of Dowd Mountain is a dip slope (a slope that conforms to the dip of the underlying rocks) on the Uinta Mountain Group. We noted no obvious changes in the rocks or vegetation.

July 11, 1968

Manila 7½' quadrangle, Utah

June 1, 1871

Red Canyon
Camera Station 534

Beaman's camera station was on the southwest bank of the Green at the immediate head of Red Canyon, close to Powell camp no. 9. Our view is upstream across Hideout Canyon, toward cliffs of Weber Sandstone that dip nearly vertically. The high cliff on the right is Uinta Mountain Group red quartzite, for which Red Canyon was named. The lower two-thirds of this cliff is now submerged; Beaman's camera station is under about 180 feet of water. Clem Powell referred in his journal to four pictures of Beehive Point taken on the morning of June 1. His "Beehive Point" was probably the more distant one.

July 11, 1968

Flaming Gorge 7½' quadrangle, Utah

June 2, 1871, early afternoon

Red Canyon
Camera Station 536

The camera station here is on the south bank of the river, below Red Canyon Overlook, about half a mile upstream from the center of the bend in the river and a third of a mile below the mouth of Eagle Creek. It is now under about 220 feet of water. View is upstream.

All rocks visible in the canyon are gently dipping beds of the Uinta Mountain Group, which consists mostly of sandstone, possibly deposited by streams, and a very few shaly intervals. We saw no noticeable changes in the cliff at the right, but great differences in perspective make a detailed comparison difficult.

36

July 12, 1968

Flaming Gorge 7½' quadrangle, Utah

June 3, 1871

Red Canyon
Camera Station 472

These pictures were taken from a station on the south bank of the Green River below Red Canyon Overlook, about three-quarters of a mile downstream from a bend in the river. View is upstream. The Beaman photograph was one of a series taken to illustrate the effect of dip on the character of canyon walls (Powell diary entry for June 3). Powell camp no. 10 was made nearby, after a portage, on the same site as Powell's 1869 camp. Beaman took six to eight pictures on the morning of June 3.

The canyon walls are all rocks of the Uinta Mountain Group. The slope near the river is not talus but a red shale unit within the Uinta Mountain sequence.

July 11, 1968

Flaming Gorge 7½' quadrangle, Utah

Ashley Falls, Red Canyon

Camera Stations 439 and 481

June 5, 1871, late afternoon

These two pictures of Beaman's show Ashley Falls and portage of the boats. His camera stations were opposite the present Cedar Springs Campground and Boat Ramp and are now under about 370 feet of water. Both pictures were taken from the north bank, which was the site of portage and of Powell camp no. 12. Photograph 439 was taken upstream from the falls, and photograph 481 was taken downstream; both show the large boulder in the middle of the falls and the boats at various stages of the portage. Stephens has duplicated the view of 481 as closely as possible.

The rocks exposed in the canyon walls are all Uinta Mountain Group.

Dutch John 7½' quadrangle, Utah

July 11, 1968

Dutch John 7½' quadrangle, Utah

Gene Shoemaker sorts through the Beaman photograph file and selects the ones we expect to encounter along the first segment of the Powell voyage. The location is near the railroad bridge, our starting point at Green River, Wyoming.

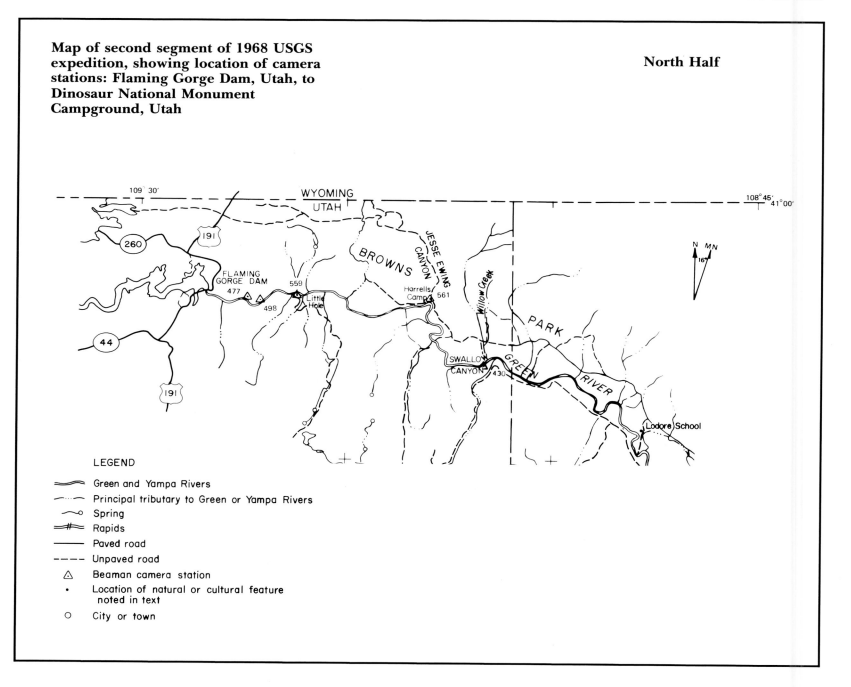

Map of second segment of 1968 USGS expedition, showing location of camera stations: Flaming Gorge Dam, Utah, to Dinosaur National Monument Campground, Utah

North Half

LEGEND

Green and Yampa Rivers
Principal tributary to Green or Yampa Rivers
Spring
Rapids
Paved road
Unpaved road
△ Beaman camera station
• Location of natural or cultural feature
  noted in text
○ City or town

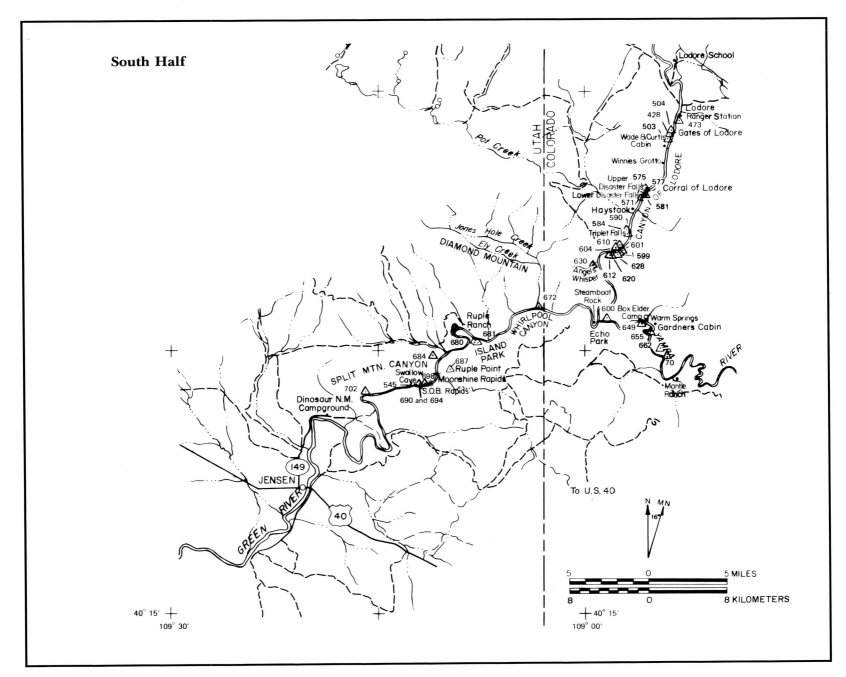

South Half

Lodore School

504
428
503
473

Lodore
Ranger Station
Gates of Lodore
Wade & Curtis
Cabin

Winnies Grotto

Pot Creek

UTAH COLORADO

Upper **575**
Disaster Falls **577**
Lower Disaster Falls
571
Haystack
590
584
Triplet Falls
610 **601**
604
630 **628**
Angels
Whisper **612** **620**

Corral of Lodore

581

CANYON OF LODORE

599

Jones Hole Creek
Ely Creek
DIAMOND MOUNTAIN

Steamboat
Rock

600 Box Elder
Camp
649
Echo
Park
655
662

Warm Springs
Gardners Cabin

672

Ruple
Ranch
681
680

WHIRLPOOL CANYON

ISLAND
PARK

70

RIVER

684
687
△ Ruple Point

Mantle
Ranch

SPLIT MTN. CANYON
Swallow
Cave **698**
Moonshine Rapids
702 **545**
S.O.B. Rapids
690 and 694

Dinosaur N.M.
Campground

To U.S. 40

JENSEN
149
40

GREEN RIVER

N MN
16°

40° 15'
109° 30'

40° 15'
109° 00'

5   0   5 MILES
8   0   8 KILOMETERS

45

Table II: Second Segment, 1968 Expedition

Geologic formations mentioned in captions of photographs taken between Flaming Gorge Dam, Utah, and Dinosaur National Monument Campground, Utah.

Geologic Era	Geologic Period	Formation or Rock Type	Remarks (Numbers are those of Beaman Photographs)
CENOZOIC	QUATERNARY	Holocene alluvium	Derived from mountain areas and deposited along river and lowland areas along much of this segment.
		Pleistocene talus	In many photographs along this segment.
	TERTIARY	Browns Park Formation	In Browns Park (561).
		Bishop Conglomerate	Caps skyline ridge of Diamond Mountain (681), Island Park.
MESOZOIC	CRETACEOUS	Mancos Shale	Island Park (681).
	JURASSIC	Morrison Formation	In distant buttes viewed from Island Park (681); forms hogback in 684.
		Stump Formation Curtis Member	Occurs near base of hogback (684).
		Entrada Sandstone	Forms low ridge in front of Morrison hogback in 684.
		Carmel Formation	In distant buttes viewed from Island Park (681).
	JURASSIC AND TRIASSIC	Glen Canyon Sandstone	In Island Park (681, 680) and 684.
	TRIASSIC	Chinle Formation Moenkopi Formation	Dark beds in 680. In Island Park (681).

Geologic Era	Geologic Period	Formation or Rock Type	Remarks (Numbers are those of Beaman Photographs)
PALEOZOIC	PERMIAN	Park City Formation	Caps Weber at Echo Park (600) and in 662 and 70.
	PENNSYLVANIAN	Weber Sandstone	Massive sandstone making up Steamboat Rock at Echo Park (600); prominent in most pictures taken in canyon of Yampa and downriver on Green.
		Morgan Formation	At Warm Springs Cedars on the Yampa (649) and in many pictures taken along the Yampa and in Split Mountain Canyon.
		Doughnut Shale	Well displayed in Split Mountain Canyon.
		Humbug Formation	Well displayed in Split Mountain Canyon.
	MISSISSIPPIAN	Madison Limestone	Caps Dunns Cliff in Canyon of Lodore (620); mouth of cave in upper Madison seen in 690, 694, 545; shown in several intervening pictures along the Green.
	CAMBRIAN	Lodore Formation	In several photographs at south end of Canyon of Lodore.
MIDDLE PROTEROZOIC		Unita Mountain Group	Resistant quartzite prominent in most pictures from Flaming Gorge Dam to below Canyon of Lodore.

Sketch of Gene Shoemaker by Stephens from a photograph taken at the end of the Canyon of Lodore segment of the voyage.

Chapter Two

Flaming Gorge Dam, Utah, to Dinosaur National Monument Campground, Utah

Eight persons in two inflatable fifteen-foot neoprene rafts began the second segment of the 1968 river trip at the foot of Flaming Gorge Dam. We reached the gravel launching ramp by a gravel road that leads down to the river. One raft was fitted with a rowing frame and oarlocks, and two of us rode in it and took turns rowing. In the second raft, six people with canoe paddles sat on the inflated gunwales and paddled according to directions from the chief boatman.

Rafting down the clear cold waters of the Green, confined within the steep red canyon walls between Flaming Gorge Dam and Little Hole, was invigorating. The few minor rapids above Little Hole provided the first "whitewater" of our trip. Fishermen often use the foot trail along the river bank to obtain excellent catches of twelve- to fifteen-inch rainbow trout. Evergreens cover the canyon walls, and growths of box elder are heavy along the river's edge and on the gravel bars.

In Little Hole, called Red Canyon Park by Powell, is an old Powell campsite on an alluvial terrace. Near the river, two large pine trees (shown in Beaman photograph 559) identify the site. The canyon widens to at least two miles as the river passes into Browns Park, where there is still road access to the favorite wintering grounds of pioneer-day cattlemen. Harrell's cattle camp of 1871 is shown in Beaman photograph 561 near several large cottonwood trees that are still alive today.

The Green River in Browns Park is nearly 500 feet wide in places, averages about three feet in depth, and follows a meandering course. Sandbars are common and we had to take care not to run the rafts onto them.

The terraced landscape in Browns Park is composed mostly of the Browns Park Formation of Tertiary age, which is dominantly sandstone, clay, and conglomerate, somewhat loosely consolidated. It imparts a buff to light-tan color to the countryside adjacent to the river. The low-lying river bank with numerous groves of cottonwood trees permits a wide choice of campsites and has plenty of firewood. Flocks of ducks and geese glided along the river ahead of our boats and took flight only when we gained on them.

Near the lower end of Browns Park, we could see the Lodore School on the east bank. Soon the Gates of Lodore, at the entrance of the Green River into the Canyon of Lodore, loomed ahead. The Lodore Ranger Station is located on the east bank about a mile upstream from the Gates of Lodore. Many boat trips begin here and end at the west edge of Dinosaur National Monument.

The most spectacular and colorful portion of the whitewater run on the Green River commences at the Gates of Lodore. It is an impressive change in scene that we shall long remember. Abruptly, the formerly slow-moving river leaves Browns Park and enters a spur of the rugged Uinta Mountains, rushing through the steep-walled canyon in a series of rapids that dealt fear to the hearts of Powell's men. Dark-red canyon walls of ancient quartzite soar 2,000 feet on either side of the river and dwarf all river runners. Disaster and Triplet falls inflicted their violence on the Powell party: one boat was upset in Disaster Falls with loss of valuable supplies. Other stretches of rapids, which forced arduous portages of their wooden boats, can now be run safely in inflatable rafts.

In the vicinity of Pot Creek—coming in from the west—light-colored Cambrian and Mississippian rocks form steep cliffs on the skyline above the red quartzite. Younger, light-colored rocks of Mississippian to Permian age are closer to river level near the junction of the Green and Yampa rivers. Echo park, at the mouth of the Yampa, is accessible by the road from the south that connects with U.S. Highway 40 about thirty-five miles east of Vernal, Utah. A National Park Service campground is located in Echo Park at the road terminus.

The most conspicuous feature at the confluence of the two rivers is Steamboat Rock—a sheer, 1,000-foot cliff of massive Weber Sandstone. Our party camped near its base on a large sandbar littered with piles of driftwood. We made a photographic sortie several miles up the Yampa to rephotograph spectacular canyon scenes captured by Beaman's camera (photographs 649 and 655).

As it flows westward from Colorado into Utah, the Green River loops around the south end of Steamboat Rock and enters Whirlpool Canyon. Faulted and folded limestones, some thin-bedded, give variety to the scenery and color to the canyon walls.

Viewed from below, the mouth of Whirlpool Canyon is a page from a geology textbook come to life: a major fault separates the intensely folded strata forming the walls of the canyon from the younger, flat-lying sandstones of Island Park downstream. The river meanders slowly through this country of low, subdued relief. Many boaters have had to row or paddle vigorously as they tediously worked their way to the head of Split Mountain Canyon against the headwinds that frequently blow up current. Major Powell applied the name "Split Mountain" to the canyon, then changed it to "Craggy Canyon," but his earlier name has been retained. Aerial views of the canyon illustrate its appropriateness.

A road from Vernal, Utah, provides access to the Green River at a launch site just above the head of Split Mountain Canyon. This site is a popular starting point for raft trips through the canyon.

Rapids of interest in Split Mountain Canyon include Moonshine and S.O.B. Downstream from them is the famous "Swallow Cave"—now devoid of swallows—named by Powell.

After running a series of rapids below the cave, our party camped on the west bank in a setting of unusual beauty. Beaman's camera station 702 on landslide debris of the canyon slope provided a spectacular view of towering sandstone crags and pinnacles towering 600 feet, silhouetted against the setting sun. When we completed the journey through Split Mountain, we returned via unpaved roads to the Yampa

River. We launched an inflatable raft on the Yampa near Mantle Ranch and floated two miles downstream to camera station 70, then floated half a mile further downstream and climbed to station 662, nearly 400 feet above the river.

Camera station 687 at Ruple Point was reached by driving over rough, unpaved roads that connect with U.S. Highway 40 to the southeast. This station, on the southeast rim of the canyon, afforded a spectacular view of Split Mountain Canyon and the Green River far below.

Above Little Hole
Camera Station 477

June 6, 1871, early afternoon

This first camera station is about three miles below Flaming Gorge Dam at a bend in the Green River. We are on the north bank of the river on a small alluvial fan that juts out into the bend, and we are looking downstream toward the next bend in the river about half a mile away.

Once our party was below the dam, the water level of the Green was much closer to that of 1871, and Stephens could duplicate the Beaman pictures more accurately. The alluvial fan can be seen in the foreground. Only the largest rock, on the left, seems not to have been moved or rotated. Others nearly as large have been either transported completely out of view or rotated enough to be unrecognizable.

July 16, 1968

Some small vegetation and a box elder now grow on the alluvial fan on which the camera station is situated.

All the rocks across the river are quartzite of the Uinta Mountain Group. No conspicuous changes are visible in them except that part of the dipping sandstone ledge in the middle distance has been blasted away to widen Little Hole Trail. Some of the largest trees along the far bank in Beaman's picture are still standing today, and they have grown noticeably.

Goslin Mountain 7½' quadrangle, Utah

June 6, 1871

Above Little Hole
Camera Station 498

We are still on the north bank, one mile below station 477 and about two-and-a-half miles above Little Hole. View is downstream. We are looking across a little rapid formed by a small talus cone of Holocene age that constricts the channel. Several boulders on the talus slope in the foreground are identifiable in both pictures. A large limb of a pine tree is balanced on the nearest boulder in the Beaman photograph. We found this limb on the river bank about twelve feet from the boulder and restored it to its approximate original position for rephotographing the scene. A juniper with many dead limbs still grows in the left foreground. A pine tree along the left edge of Beaman's picture is present today, and it looks very much the same now as in 1871. On the right bank, three pine trees can still be seen in the distance; two had rounded crowns in 1871. The dead pines in the middle distance on the right side of the 1871 picture have fallen.

In other respects, the right bank has changed considerably. The beautiful sand beach below the rapid in the Beaman picture has been largely washed out and replaced by an eddy. Its erosion may be due to the construction of Flaming Gorge Dam and the consequent decrease in sand deposition after the high spring runoff. Box elders are encroaching on the edge of the old sand beach. All the rocks shown are quartzite of the Uinta Mountain Group. We detected no changes in the bedrock.

July 16, 1968

Goslin Mountain 7½' quadrangle, Utah

June 7, 1871, morning

Little Hole
Camera Station 559

The camera station is in Little Hole on an alluvial terrace, probably of Holocene age, in the center of the park. We are on the south bank of the river, about ninety feet east of Powell's camp no. 13 and about fifty feet from the water's edge. The view is upriver. Beaman took this picture on a day when most of the Powell party were climbing the nearby mountains to study the geology and sketch the topography, but we can see Hillers seated near one of the large pine trees and several other men between the trees and the tent on the left.

At the extreme left side of the pictures is a bluff at the edge of a Pleistocene terrace. What appears to be a ledge on the bluff in the Beaman photograph is the wall of a small gully cutting the terrace. The gravels of the upper terrace have since been excavated, possibly for road metal or placer gold. In the distance are rounded hills of quartzite of the Uinta Mountain Group.

The two large pine trees in the right foreground are not much different today than when Powell camped under them. A pine that stood in the distance next to some dead trees is now gone. Young pines have grown up on the skyline ridge since Beaman's picture was taken, and one old-timer is still there.

July 16, 1968

Goslin Mountain 7½' quadrangle, Utah

June 1871

Browns Park
Camera Station 561

Beaman took this photograph about half a mile up the Green from its junction with a road from Jesse Ewing Canyon. The camera station is on the north bank of the river about seventy-five feet from its edge. We are looking southwest, upriver.

Beaman's picture shows Harrell's camp, a cattle station that had occupied the site intermittently since before the 1870s. Powell's camp no. 14 was evidently close to the camera station, as a large cottonwood tree described by Dellenbaugh (in *A Canyon Voyage*) still grows just to the left of the scene.

This site was also a camp in 1968. The principal differences are that the covered wagons of 1871 have been replaced by modern house trailers, and a vegetable garden occupied the foreground area. Two log cabins of ancient vintage are obscured by the larger trailer. A third cabin to the left of the scenes might be Christian Hillman's cabin mentioned in Powell's journal. Many cottonwood trees visible in Beaman's picture are still alive, although the one on the right is just barely holding on. Junipers on the bluff on the opposite bank of the river, seen through the gaps in the cottonwood trees, still occupy the same sites, although it is difficult to identify individual trees.

The low bluff forming the far river bank is capped by Pleistocene terrace gravel. Beneath the gravel are light-colored outcrops of the Browns Park Formation, which can just be discerned between the trees in Beaman's photograph. Two successively higher ridges in the distance are formed of quartzite of the Uinta Mountain Group. The foreground is a terrace of Holocene alluvium.

July 17, 1968

Clay Basin 7½' quadrangle, Utah

Swallow Canyon (Browns Park)
Camera Station 430

June 1871

These pictures were taken at the mouth of Swallow Canyon, a third of a mile above the mouth of Willow Creek. The station is on the south bank of the river about a hundred feet upstream from the east end of low cliffs that flank the river. The view is west (upstream) toward a salient of the Uinta Mountains that extends into Browns Park from the south. The salient is hard quartzite of the Uinta Mountain Group, but the Green River was able to cut through it, forming Swallow Canyon, because it had originally established its position in the overlying softer Browns Park Formation, now stripped away. All bedrock in the pictures is quartzite of the Uinta Mountain Group.

The water level in the 1871 picture seems unusually low for the time of year, although it may have dropped a foot or two after the party entered the canyon. The foreground rocks in the two pictures

July 17, 1968

are nearly identical, but the difference in lighting makes detailed comparison difficult.

Beaman probably took this picture in the morning or possibly in the early afternoon. There is some doubt about its origin. Powell's journal says that they "camped for dinner on the rocks near foot of the [Swallow] canyon in the shade of a cliff while Beaman was taking pictures above [the cliff]." Del-lenbaugh published the picture (erroneously labeled "Red Canyon Park") in *A Canyon Voyage* and attributed it to Beaman. Beaman probably took it, but the site also could have been easily reached by W. H. Jackson or later by Jack Hillers. A handwritten note in the Hillers album says "Photo by W. H. Jackson, 333, 1870," but many photographs have been incorrectly attributed to Jackson.

Swallow Canyon 7½′ quadrangle, Utah

Canyon of Lodore
Camera Station 473

June 17, 1871, morning

We are now well into Colorado, more than twenty river miles downstream from the previous camera station. We are looking downstream, and the river is flowing generally south-southwest. The camera station, on the east bank of the river at the Gates of Lodore, is about 200 feet from the probable location of Powell camp no. 16. Powell's men rowed their boats across the river so they would appear in Beaman's picture, and we can see the three tiny boats near its center.

Box elder trees are growing today at the site of the two prominent trees in Beaman's foreground, and we saw many young ones where only low shrubs and sand appear in the Beaman photograph. Because the campsite was less congested with vegetation in Powell's day, it was more practical for his party than for ours.

The general configuration of the sandstone ledge in the immediate foreground is today nearly identical with that of 1871, although a few loose rocks seem to have been moved. Canyon walls in the distance are quartzite of the Uinta Mountain Group.

July 18, 1968

Canyon of Lodore North 7½' quadrangle, Colorado

June 17, 1871, morning

Canyon of Lodore
Camera Station 504

The camera station is on the west bank of the Green about one and a quarter miles below Powell's camp no. 16 and one mile below the Gates of Lodore. The view is east, across the river. The exact station could not be occupied because it is now covered by a boulder whose top is about twelve feet across, one of several huge blocks that have tumbled onto the site from the high canyon wall during the past century.

All the rocks in view on the opposite side of the river are quartzite of the Uinta Mountain Group. A steep alluvial fan extends in a broad curving arc from the mouth of a steep-walled gully to the river. The main part of the fan is probably of Pleistocene age, but it is still active. We detected no changes in the bedrock or fan, except that the conspicuous stripe at the head of the fan, shown in Beaman's picture, is no longer present. More juniper and piñon now grow on the slope on the left side of the picture, and pine or Douglas fir have gained a foothold along a rocky defile high on the right side.

64

July 19, 1968

Canyon of Lodore North 7½' quadrangle, Colorado

Canyon of Lodore
Camera Station 428

June 17, 1871, morning

We are here on the west bank of the river about 200 feet downstream from the last camera station, looking downstream. We had not expected to find the same sandbar, but there it was! The lagoon in the foreground has been partly filled in and is now a horsetail marsh. The other principal change is the invasion of the river bank by tamarisk. The man in the Beaman picture is Dellenbaugh. The dead branch near him may be the limb of the large cottonwood whose trunk is still on the river bank just out of view of our picture. Juniper, piñon, and box elder line both river banks today as they did in Beaman's time. The canyon walls are all quartzite of the Uinta Mountain Group.

July 19, 1968

Canyon of Lodore North 7½' quadrangle, Colorado

June 17, 1871, midafternoon

Canyon of Lodore
Camera Station 503

The camera station is on the west bank of the river, about one-and-a-third miles below the Gates of Lodore. The view is across the river to the southeast. The water level was one-and-a-half to two feet lower at the time of Beaman's picture, but many of the larger rocks visible in the river of 1871 were also projecting above water when we were there.

The rocks of the canyon wall are quartzite of the Uinta Mountain Group. In the gully near the center of the pictures, a still-active Pleistocene talus cone (in shadow in Beaman's photograph) sheds a steep, coarse alluvial fan toward the river and is the principal feature of the pictures. Box elders and junipers on Beaman's far bank are in about the same positions today.

July 19, 1968

Canyon of Lodore North 7½' quadrangle, Colorado

69

June 18, 1871, afternoon

Canyon of Lodore
Camera Station 577

The camera station is at the head of Upper Disaster Falls on the northwest bank of the river. View is south across the river at a promontory west of the Corral of Lodore. The station was located with reference to rocks in the left foreground. The sand that partly covered these rocks has been shifted by the wind, exposing more of some rocks and less of others. The two largest junipers on the left side of the 1871 picture and the mountain mahogany bush between them have scarcely changed shape, but the three-foot piñon tree just to the left of the figure in the Beaman picture shows the growth one would expect. The carcasses of junipers to the right of the figure have nearly all of their earlier branches.

Across the river is a talus cone grading out into an alluvial fan. This fan and one opposite, beneath the camera station, constrict the river here, and large blocks that have been freed from the fans are responsible for the rapids. The cones and fans are probably Pleistocene in age, but the talus at their heads is active. The cone across the river heads into the gully of the Corral, which may mark the trace of a fault (as suggested by P. T. Hayes, U.S. Geological Survey). The rocks exposed are quartzite of the Uinta Mountain Group.

July 20, 1968

Canyon of Lodore North 7½' quadrangle, Colorado

June 18, 1871, afternoon

Canyon of Lodore
Camera Station 575

The camera station is at Upper Disaster Falls on the west bank of the river, and the view is downstream. The old juniper carcass in Beaman's foreground is still standing, but younger junipers have grown into adult trees and now obscure the view in the center of our picture.

All the rocks in view are quartzite of the Uinta Mountain Group. A fairly large fan extends from the foot of the canyon wall on the left; the fan lies at the mouth of drainage from a substantial amphitheater. This fan crowds the river against the west canyon wall and is responsible for the lower part of Upper Disaster Falls.

July 20, 1968

Canyon of Lodore North 7½' quadrangle, Colorado

Canyon of Lodore
Camera Station 571

June 18, 1871, afternoon

This camera station is on the west bank of the Green 200 feet downstream from station 577. View is across the river to the southeast, up the gully of the Corral of Lodore. Two piñons have grown up to frame the twisted juniper limb seen in both pictures. A young piñon at the extreme left of Beaman's photograph is now an adult tree.

The features across the river are virtually the same as those described for Beaman photograph 577.

74

July 20, 1968

Canyon of Lodore North 7½' quadrangle, Colorado

June 19, 1871, early afternoon

Canyon of Lodore
Camera Station 581

The scene here is of the lower end of Upper Disaster Falls, taken from the east bank of the river. View is upstream, showing some of the larger rapids in the falls. Our camera station was located within a few feet of the original site, but no foreground features or rocks in the river could be correlated with those in Beaman's picture. Comparison of the far canyon walls shows that the water was as much as three feet higher at the time of Beaman's picture than at the time of ours. This higher water level suprised us: the Beaman photographs taken on previous days show lower water than do ours. The water must have risen suddenly between June 16 and 19, 1871.

The far bank of today appears identical with that shown in Beaman's picture. Even relatively small carcasses of juniper trees are still there, and small rocks appear to be in identical positions.

All rocks exposed are quartzite of the Uinta Mountain Group. Just above the river on the right side of the picture is the Pleistocene fan on the northwest bank that is partly responsible for the rapid at the head of Upper Disaster Falls.

July 20, 1968

Canyon of Lodore North 7½′ quadrangle, Colorado

June 21, 1871, morning

Canyon of Lodore
Camera Station 590

We are here on the west bank of the Green about two-thirds of a mile below the mouth of Pot Creek (called Cascade by Major Powell). View is downstream. Thompson wrote: "Stayed in camp while Beaman took pictures until about 10:00 a.m." The boats pulled up on the bank downstream probably mark the site of the noon camp of June 21.

Beaman's camera station was located precisely by relating it to the two large rocks at the edge of the water. These rocks, partly under water in Beaman's picture, were high and dry when we saw them. The difference in water stage, as at the previous station, is about three feet. Many rocks projecting above the sand in Beaman's picture are more deeply or completely buried today. One fresh, jagged rock has tumbled into the center of the scene, apparently from the cliffs on the right. The pine tree in the middle distance in the center of the pictures has grown at least thirty feet. The deciduous tree to its right, perhaps a very large box elder, is gone. Along the skyline on the opposite side of the river some trees are still standing, but others are gone and new trees have grown up.

The cliff in the middle distance is quartzite of the Uinta Mountain Group. Beyond it are two cliffs, one right behind the other, which are hard to distinguish in Beaman's picture. The lower, light-colored cliff is Lodore Formation of Cambrian age and the upper cliff is Mississippian limestone.

July 21, 1968

Canyon of Lodore South 7½' quadrangle, Colorado

79

June 21, 1871, morning

Canyon of Lodore
Camera Station 584

The camera station is 600 feet downstream from camera station 590, on the west bank of the river on the point of an alluvial fan. View is upstream of the Haystack (called Wheat stack by the Powell party), which lies just north of the mouth of Pot Creek. The exact location of the camera station was determined by the position of the large rocks in the foreground of Beaman's picture, but the presence of a piñon tree squarely in the center of the view required that Stephens's picture be offset to the left. Other trees are mostly box elders. In the 1871 picture, a large rock in the middle of the river, beneath the dark cliff on the right, is nearly covered by water. This rock was four feet out of the water when we were there.

All of the rocks in view are quartzite of the Uinta Mountain Group.

July 21, 1968

Canyon of Lodore South 7½' quadrangle, Colorado

June 21, 1871

Canyon of Lodore
Camera Station 610

This camera station is at the same west-bank location as that of the next station (604), but the view is downstream toward the second rapid in Triplet Falls. The water was about ten feet closer to the station in 1871 than in 1968. Rocks on the opposite shore lie at the foot of a colluvial fan that crowds the river against the west wall of the canyon, forming the falls. Piñon and juniper growth on the fan appear to be about the same today as in 1871.

The middle and lower canyon walls are quartzite of the Uinta Mountain Group. The Lodore Formation caps the cliff on the left side of the pictures and forms the promontory framed by the canyon walls at upper right. A slope of Mississippian-age rocks can be seen in the extreme upper left corner of the pictures.

82

July 22, 1968

Canyon of Lodore South 7½' quadrangle, Colorado

Canyon of Lodore
Camera Station 604

This camera station, like the previous one, is on the west bank of the river at the head of Triplet Falls. The view is of the first rapid and, across the river to the southeast up a side canyon, of the eroded fan that is responsible for the rapids. One small fan directly across the river has contributed some of the boulders that form the first rapid; a small fan opposite it underlies the camera station. Drainage from the side canyon in the center of the pictures bends to the right and joins the river at the right side of the pictures just above a small beach. The water here is three-and-a-half to four feet higher in the 1871 picture than in ours. The exact camera station was located in relation to foreground rocks, but Stephens's picture is offset about ten feet to the right because the view was obstructed by a box elder. Box elders on the small fan across the river have also grown considerably since 1871.

Most of the opposite canyon wall is quartzite of the Uinta Mountain Group. The Lodore Formation forms the upper part of the promontory on the right skyline, the relatively smooth cliff just above the more rugged quartzite, and the tree-covered slope. Above the Lodore, Mississippian limestone forms the cliffs of the saddle at the skyline.

June 21, 1871

84

July 22, 1968

Canyon of Lodore South 7½' quadrangle, Colorado

June 22, 1871, 10:00 A.M.

Canyon of Lodore
Camera Station 601

These pictures were taken at Triplet Falls from the east bank of the river. The view is downstream, across the lowermost of the three rapids. Our camera station was located precisely in relation to the pile of foreground boulders, which today is obscured by box elders. Almost all the rocks are in exactly the same positions today as when Beaman photographed them, although the water was then four to five feet higher and covered more of them.

Most of the foreground boulders are dark-reddish-brown quartzite of the Uinta Mountain Group, with a few light-gray limestone boulders, probably from rocks of Mississippian age, scattered among them. The darker cliffs in the middle distance on both sides of the river are Uinta Mountain quartzite. The sunlit cliff along the skyline, framed by the nearer canyon walls, is Lodore Formation.

Most features of the canyon walls today are identical with those in the Beaman photograph. However, one light area in our picture, high on the canyon wall slightly to the right of center, may be the source of a post-1871 rockslide. Near the river, the drapery patterns of desert varnish on the cliff to the left of center appear identical today with those shown in Beaman's picture. Prominent Douglas fir and pine trees along the skyline occupy the same positions but have changed shape. The juniper framed by the overhang on the upper right canyon wall has grown considerably.

The figure reclining against a rock at the extreme left in the 1871 picture is probably Clem Powell. His journal for June 22 says, "The Major, Professor, and Captain started out for an all day's climb while Beaman and myself went down to the river to take views of as grand scenery as the sun ever shone on. Took seven views of Triplet Falls and vicinity, I being in some of them."

July 22, 1968

Canyon of Lodore South 7½' quadrangle, Colorado

87

June 22, 1871, about 10:00 A.M.

Canyon of Lodore
Camera Station 599

The camera station is at Triplet Falls on the east bank, about 160 feet downstream from the previous station. We are looking directly across the river at the canyon wall just below the lowermost of the three rapids. The large rock in the water on the right is the same rock in the water on the right side of Beaman photograph 601, but here it is seen at an angle of nearly 90° from the line of sight in 601. The three largest rocks in the water were mostly exposed in our picture, as the water was lower and the main part of the flow was farther out toward the center of the river. Stephens's photograph was taken slightly later in the day than Beaman's, but the shadow patterns almost match.

The exposed rocks are Uinta Mountain Group quartzite dipping gently downstream. The stratum under the overhanging ledge may be the source of some of the large blocks in the river, as the wall behind is devoid of desert varnish and appears rather fresh. The wall on the extreme left is highly stained and is evidently the site of an intermittent seep. Most of the stains appear to be unchanged, although the high contrast of the Beaman print makes detailed comparison difficult. Small box elders have sprung up at the foot of the cliff in an area that was partly under water at the time of the Beaman photograph.

July 22, 1968

Canyon of Lodore South 7½' Quadrangle, Colorado

June 22, 1871, late morning

Canyon of Lodore
Camera Station 628

We are here just below Triplet Falls on the south bank of the river, looking downstream. The standing wave tail of the lowermost rapid of Triplet Falls can be seen on the right. The exact camera station location is related to a large foreground rock, largely hidden by sagebrush in the Beaman photograph. This pair of pictures is one of the few that shows less foreground vegetation today than in 1871. The amount of sagebrush has declined, but box elders have grown up and now frame our view of the river.

The canyon walls on either side of the river are chiefly Uinta Mountain Group quartzite. The cliff capping the promontory in the center of the pictures, framed by the canyon walls, is Lodore Formation, as is the smooth cliff on the skyline in the upper left corner.

July 22, 1968

Canyon of Lodore South 7½' quadrangle, Colorado

June 23, 1871, afternoon

Canyon of Lodore
Camera Station 620

The camera station is on the south bank of the river just above Boulder Falls at the head of Hell's Half Mile. View is east (upriver) toward Dunns Cliff, framed between the canyon walls. The 1871 picture was probably taken during the beginning of the portage around Hell's Half Mile. The foreground bank was the site of Powell's camp no. 22. At the head of the portage, it is a convenient camp even today. Shade is provided by large box elders. Descendants of the serviceberry and juniper in the foreground of Beaman's picture are growing on the site today.

The lower canyon walls are quartzite of the Uinta Mountain Group. Dunns Cliff is capped by Madison Limestone of Mississippian age. The tree-covered slope beneath it is the lower part of the Madison. Below the Madison is a bare cliff of Lodore Formation resting unconformably on the Uinta Mountain Group. (In the opposite direction, downriver, we could see the unconformity at the base of the Lodore cutting out beds of the Uinta Mountain Group at an apparent angle of about 5°.)

July 22, 1968

Canyon of Lodore South 7½' quadrangle, Colorado

June 23, 1871, afternoon

Canyon of Lodore
Camera Station 612

We are standing on the south bank of the river at Boulder Falls, at the head of Hell's Half Mile. View is directly across the falls and up a side canyon (named Boulder Gulch by Thompson), whose eroded fan has contributed the boulders on the far side of the river. The boulders on the near bank are from a matching fan at a canyon mouth on our side of the river. This fan is chiefly responsible for the falls. Both tributary canyons follow a north-south fault in the Uinta Mountain Group, which is the rock in view here. The fault is just to the right of the small isolated butte at the canyon mouth in the center of the photographs. West (left) of this fault, the rocks dip 25° to the west.

Beaman's picture shows Jones, Hillers, and Dellenbaugh at work lining one of the boats. In 1968 the foreground backwater in which the men had been standing was high and dry and grown over with sweet clover, horsetail, junipers, and scattered small tamarisks. Many more boulders were also exposed in the river channel. The large cottonwood log perched on the big foreground boulders in 1968 may be the large snag seen a few feet upstream in the Beaman photograph. Several small box elders have sprung up on the far bank since 1871.

July 22, 1968

Canyon of Lodore South 7½' quadrangle, Colorado

June 24, 1871

Canyon of Lodore
Camera Station 630

The camera station is on the southwest bank of the Green near the mouth of Rippling Brook (named Leaping Brook by Major Powell). View is west, of the main tributary of the brook. Powell named this tributary the Angel's Whisper because of the soft whispering sound of the multiple cascades of a tiny, but evidently permanent, stream. Would we today choose such a name?

The lower parts of the canyon walls are of Uinta Mountain Group quartzite. The uppermost cliffs in the middle ground of the pictures are Lodore Formation. At the head of the canyon, framed by the canyon walls, is the Madison Limestone, more easily recognized in the 1871 picture where it is sunlit. No geologic changes were detected in the canyon walls.

A fine stand of Douglas fir follows the watercourse along the Angel's Whisper. Scattered among the dark conifers are smaller deciduous trees and shrubs of brighter green, which stand in striking contrast to the dark-red rocks of the Uinta Mountain Group. Several of the trees shown in Beaman's picture can be recognized, particularly those nearest the camera; two trees flanking the lower cascade have grown twenty to twenty-five feet.

The camera station, several hundred feet above the Green River, is a spot well worth visiting. It is a tribute to

Powell's men that they succeeded in carrying Beaman's awkward, heavy photographic equipment to it. Our party needed two hours of hard scrambling to locate the station.

July 23, 1968

Canyon of Lodore South 7½' quadrangle, Colorado

July 1, 1871, morning

Echo Park
Camera Station 600

We are now in Echo Park, on a bench of sandstone between the Yampa and Green rivers. View is to the southwest, toward the junction of the rivers and in the downstream direction of the Green. The Yampa flows in from the left. Steamboat Rock juts out in the right side of the pictures. Powell's camp no. 24 was probably in the box elder grove near the foreground bank of the Green. The two pictures were taken at nearly the same time of day.

The island in the center of the pictures has changed considerably. It now extends much farther upstream in the Green but has not blocked the channel between the Green and the Yampa in the foreground. The left bank of the Yampa, which in 1871 was alluvial fill supporting box elders, has been cut back to a bedrock sandstone ledge. The foreground ledge of sandstone has scarcely changed.

Sandstone in the lower part of the Weber Sandstone of Pennsylvanian age forms the foreground rocks, almost all of Steamboat Rock, and the canyon wall in the middle distance. These high cliffs are capped by a very thin, dark layer of Park City Formation of Permian age. In the distance we can see the ridge of Pearl Park, which is also Weber capped by Park City. Between the middle distance and the far ridge, these rocks were squeezed into two long

98

folds whose axes paralleled the face of the ridge. Erosion has removed part of the intervening rocks, leaving the high ridge of Pearl Park.

No changes were detected in the bedrock of the distant scene. Even the pattern of dark drapery on Steamboat Rock appears the same.

July 24, 1968

Canyon of Lodore South 7½' quadrangle, Colorado

June 27, 1871, late afternoon

Yampa River Canyon
Camera Station 649

These pictures were taken from the south bank of the Yampa River at the foot of Warm Springs Cliff, directly opposite the mouth of Warm Springs Draw. (We found a good flow of water from springs in the area.) The view is across and upriver (to the east), showing the promontory behind Gardners Cabin. An open, park-like area on the opposite side of the river, out of sight to the left, was named Grizzly Park by Powell but is called Warm Springs Cedars today.

The promontory is composed of the upper part of the Morgan Formation of Pennsylvanian age. The peak just behind and to its left is Weber Sandstone, which also forms the distant bright cliff framed by the canyon walls. The base of the Weber is a little below the alcove framed by the shadow and the box elders. The visible part of the dark cliff on the right is Morgan Formation.

This camera station is the site of one of the most sudden, dramatic changes in the canyons that human beings have ever witnessed. One night in early June 1965, some campers at Warm Springs were alarmed by a flash flood in Warm Springs Draw. It deposited a fan of very coarse boulders on the right bank of the draw, just upstream from its mouth. This fan forced the Yampa against its left bank at the foot of Warm

Springs Cliff and completely changed the character of Warm Springs Rapid. The following day, a professional boat-man lost his life in the transformed rapids.

The river has since cut away part of the fan and scoured out the left bank. All the foreground rocks are gone, including the huge one behind Powell's chair. The exact camera site could not be occupied because it is now in the river, about ten feet from the bank. Stephens's photograph shows part of the recent fan across the river.

No obvious changes were observed in the distant background.

July 24, 1968

Canyon of Lodore South 7½' quadrangle, Colorado

Yampa River Canyon
Camera Station 655

June 28, 1871, afternoon

The camera station is about one-and-a-half miles above Warm Springs, on a point within a bend of the Yampa. The view is downstream, of a wall of Weber Sandstone and Morgan Formation. The knob on the skyline at the upper left is the same one framed by the canyon walls in Beaman photograph 649, but here it is viewed from upriver.

This camera station was reached with some difficulty. We (Stephens and Shoemaker) went upriver from Echo Park in a small john boat. At Warm Springs Rapid we left the boat, located camera station 649, and then hiked upstream on the south bank. Stephens swam across the Yampa to station 655 and took his photograph in the fading twilight. We did not have sufficient time to determine the precise location of Beaman's station, which is actually located farther downstream, near the rocks on the right bank seen in Stephens's picture. Then we floated back down to Warm Springs Rapid on a large cottonwood log, scrambled over rocks and precipitous ledges in total darkness, located the boat, and proceeded downriver. We arrived in camp at Echo Park just after midnight.

The pictures show a wall of Weber Sandstone overlying ledge-forming slopes of the Morgan Formation. The contact of the two formations is about

102

one-third of the vertical distance up from the river. The Morgan also makes up the ledge of interbedded sandstone and limestone in the banks of the middle distance, but it is covered by talus on the right bank. The beds dip gently toward the camera station.

July 24, 1968

Canyon of Lodore South 7½' quadrangle, Colorado

June 29, 1871, early afternoon

Yampa River Canyon
Camera Station 662

Our station here is above the right (northeast) wall of Yampa River Canyon, about two-and-a-half miles below Mantle Ranch and 400 feet above the river. The view is downcanyon to the northwest. The exact camera station location was based on parallax in the middle distance of Beaman's picture, a tree in the lower right foreground, and small sandstone points on the rim of the butte supporting the camera station.

The light-colored sandstone point in the lower center of the Beaman picture is now obscured by a small shrub. The pattern of Mormon tea growing in the foreground, similar to that in 1871, suggests that the bushes are the same. An old juniper log at the lower right edge of the Beaman picture is still present, although more weathered, but a juniper snag near the butte's rim on the left side of the picture is now toppled over. A few deer trails that stripe the sandy colluvium in the middle distance are the only other conspicuous differences between the pictures.

Most of the rocks that we see are Weber Sandstone. Beds of the Park City Formation cap the canyon walls on either side, but here they are not more than fifty feet thick. The uppermost beds of the Morgan Formation are exposed at the base of the cliff near river level, almost squarely in the

center of the pictures. They form subdued ledges dotted with small junipers.
Some of the lines of prominent junipers
in the middle distance follow well-developed joints in the sandstone. The
butte on which the camera station is
located is capped by terrace gravel,
probably of early Pleistocene age.

July 30, 1969

Canyon of Lodore South 7½' quadrangle, Colorado

June 29, 1871, morning

Yampa River Canyon
Camera Station 70

This camera station, about eight miles up the Yampa from Echo Park, is near the mouth of a side drainage entering from the north about two miles below Mantle Ranch. View is downstream to the west. Although the foreground cleft framing the scene appears narrow in the pictures, it is in fact about 200 feet across. The exact camera station was based on parallax between foreground cliffs and the distant features downriver. Stephens took his photograph in the afternoon. The river was one to two feet higher at the time Beaman's picture was taken than it was at the time of ours.

A juniper tree close to the camera on the left side of Beaman's picture may be the same one that we see today. Box elders still line the river bank at the foot of the canyon.

Most of the canyon walls in view are the Weber Sandstone. The uppermost beds of the Morgan Formation form ledges at the foot of the side canyon and along the river bank in the middle distance. The Park City Formation caps the bluff in the center of the pictures and a more distant bluff to the right. A fracture or shear zone (fracture accompanied by deformation) separates these two bluffs; because the rocks in this zone have been weakened, erosion has formed the notch on the horizon. The beds are dipping gently downstream, away from the camera. The height of the bluff in the center of the picture is almost one thousand feet. No conspicuous changes were found in the foreground cliffs.

July 30, 1968

Hells Canyon 7½' quadrangle, Colorado

107

Whirlpool Canyon

Camera Station 672

We are now back in Utah, again on the Green River. The camera station is on Jones Hole Creek (called Bishop's Creek by J. W. Powell) about two hundred feet from its mouth. The view is northwest across the creek, toward the corner of the canyon wall that separates the canyon of Jones Hole Creek from Whirlpool Canyon. Beaman's precise camera station could not be located because of major changes in the foreground. His picture shows the stream flowing almost directly toward him, away from the massive cliffs. In 1968, the stream flowed more southerly, in almost a straight line to the Green River, from right to left across the field of view. Apparently it has cut down as well as straightened its course; we could not estimate the height of the bank in Beaman's photograph because of the dense growth of box elder. The Madison Limestone of Mississippian age forms the high cliff and an underlying series of limestone ledges and greenish-gray shale interbeds.

July 4, 1871, 9 A.M.

July 27, 1968

Jones Hole 7½' quadrangle, Utah-Colorado

July 6, 1871, afternoon

Island Park
Camera Station 681

This camera station is above the south bank of the Green, about a mile upriver from Ruple Ranch. The view is downstream (to the northwest), showing a major group of islands in Island Park a few hundred yards below the mouth of Whirlpool Canyon. The islands have changed considerably since 1871. The one on the extreme right in Beaman's photograph now extends past the middle of the 1968 picture and today is called Buck Island. It has grown by major additions of sediment on the left (downstream) side. The most distant island, covered with cottonwood trees in Beaman's picture, is Big Island. By 1968 it had lost considerable ground on the downstream side but had gained on the upstream end. The main channel of the river now cuts slightly to the left of the center of our photograph. The two nearest islands in the 1871 picture, one with a substantial patch of cottonwoods, have disappeared.

In Beaman's photograph, two cottonwood trees are rooted in a lower level of alluvium at the foot of a terrace, their heads showing above the terrace and the Glen Canyon bluff. Two cottonwoods are growing in this position today; the one on the right is probably the same as that in Beaman's photograph. Another cottonwood has grown up from the foot of the terrace near the right edge of the 1968 picture. In the left foreground, a juniper silhouetted against the river in Beaman's photograph is now a carcass.

In the right foreground, mountain mahogany brush obscures a terrace of Holocene alluvium. In the left foreground is a low bluff of Glen Canyon Sandstone, partly covered by sand derived from the Glen Canyon. The bluffs in the middle distance on both sides of the river are also Glen Canyon Sandstone, capped by Pleistocene terrace gravels. Glen Canyon Sandstone also forms the buttes to the right in front of the skyline ridge. Farther downstream, the buttes in the haze near the center of Beaman's photograph consist of the Carmel to Morrison Formations dipping about 10° to the left (southwest). However, remnants of early Pleistocene sediments cap part of these bluffs on the left, obscuring the Carmel. The skyline ridge is Diamond Mountain, capped on the left by the Bishop Conglomerate underlain by Mancos Shale, which dips generally to the left. The Moenkopi Formation is exposed on the extreme right edge of Diamond Mountain.

To take this picture and the following one, Beaman, Dellenbaugh, and W. C. Powell rowed four miles upriver from their camp at the head of Split Mountain Canyon.

July 27, 1968

Island Park 7½' quadrangle, Utah

111

Island Park
Camera Station 680

July 6, 1871

This camera station is 120 feet southwest of station 681, also on the south bank of the Green. View is southeast, toward the south wall of Whirlpool Canyon and of a small subsidiary canyon roughly parallel to Whirlpool Canyon.

All the rocks in view are Weber Sandstone except for some dark beds of Chinle Formation in the lower right and a few outcrops of Glen Canyon Sandstone in the foreground. The subsidiary canyon has developed along a fault that has dropped the Chinle and Glen Canyon to the level of the Weber. The trace of the fault is almost perpendicular to the camera line of sight. In the middle distance, the beds are dipping steeply toward the camera, but the dip flattens both upcanyon and in the foreground. Most of the foreground and the ridges in the middle distance are covered by sand derived from the Glen Canyon.

The pattern of trees in the middle and far distance is almost identical in the two pictures. Most of the trees in the canyon are Douglas fir; a few box elders are growing today at its mouth. The pattern of junipers in the foreground is somewhat different today, but a few living trees can be correlated with those of 1871.

July 27, 1968

Island Park 7½' quadrangle, Utah

113

July 8, 1871, morning

Split Mountain Canyon
Camera Station 687

Beaman probably reached this site at Ruple Point by way of Island Park, because the site is separated from the river by nearly vertical cliffs. We reached the station by unpaved road south of Split Mountain.

The view is downstream, to the southwest. The character of the rapids is similar in both pictures. Moonshine Rapids are closest to the camera. S.O.B. Rapids are in the center of the left half of the pictures between two tight bends of the river, and Schoolboy Rapids are just beyond. This view includes three camera stations that Beaman occupied two days later: station 698 (at the foot of S.O.B. Rapids) and stations 690 and 694 (at the head of Schoolboy Rapids).

The ledges in the left foreground and the rocks forming the ragged skyline are light-tan and cream-colored, massive, crossbedded Weber Sandstone. Weber also caps the ridges on each side of the river. Below the Weber, thick-bedded limestone, sandstone, and a few thin shale beds of the Morgan Formation make up the distinctive red cliffs of the middle third of the canyon walls. The slopes along the river are composed of dark shale of the Doughnut Shale and the few cliffs along the river are massive gray limestone of the Humbug Formation. Across the river on the west bank are large cones of Pleistocene talus and colluvium.

July 30, 1968

Split Mountain 7½' quadrangle, Utah

July 8, 1871, afternoon

Little Rainbow Park
Camera Station 684

We are now in Little Rainbow Park on the west side of the Green, about half a mile from the river. The view is east-northeast toward the mouth of Whirlpool Canyon, which shows dimly in the distant haze of Beaman's photograph. The river and the north flank of Split Mountain are on the extreme right side of the scene. Beaman took his picture after he had climbed down from Ruple Point, from which he had taken the previous picture of Split Mountain Canyon. We located his camera station exactly on the basis of dipping foreground beds of Glen Canyon Sandstone. An old juniper carcass below us and slightly right of center is still rooted in the bedrock.

Rocks in the pictures range from the Morgan Formation in Whirlpool Canyon to the Morrison Formation, which forms a subdued, brush-covered hogback in the middle distance that extends from the left across the middle of the pictures. The low, dissected ridge that obscures most of the Morrison hogback is Entrada Sandstone; the right end of the ridge is covered by Pleistocene gravel on which junipers are growing. Faulting and folding have placed the older Weber Sandstone at higher elevations along the flank of Split Mountain. A well-developed Pleistocene pediment (the sloping erosional surface at the base of the moun-

116

tain) extends from this flank toward the river. The valley of Little Rainbow Park on the near side of the river is floored by Holocene alluvium.

The configuration of the foreground ledges of Glen Canyon Sandstone is nearly the same today as in 1871, but small changes can be detected in the lower ledge of the projecting point. More box elders line the river bank today and more junipers are growing at the end of the Entrada ridge.

July 28, 1968

Split Mountain 7½' quadrangle, Utah

Split Mountain Canyon
Camera Station 698

July 10, 1871, morning

The camera station is on the north bank of Green River at the foot of S.O.B. Rapids. View is downstream to the west-southwest. The Beaman picture shows the boats *Nellie Powell* and *Cañonita* tied to the rocks and duffle stacked on the shore. Major Powell and Jack Hillers had left the party that day and hiked ahead to the Uinta Indian Agency.

We had trouble locating the precise camera station because the positions of foreground boulders have changed. The estimated precision of location is about five feet. A snaggy old cottonwood on the right side of Beaman's photograph may be represented by an overturned log with roots attached that lies in approximately the same position today (just out of view). What appears to be a young Douglas fir near the center of the 1871 picture is probably the large tree that stands on the site today.

The opposite river bank is a deeply dissected Pleistocene fan, which is red because it is rich in detritus from the red rocks of the Morgan Formation. The shapes of the lowermost part of the fan and of the accumulation of boulders at the water's edge have changed considerably. Bedrock exposed on the slopes above the fan includes the Humbug Formation, Doughnut Shale, and Morgan Formation. The skyline crags are Weber Sandstone.

July 28, 1968

Split Mountain 7½' quadrangle, Utah

July 10, 1871, morning

Split Mountain Canyon
Camera Station 690

We are still on the north bank of the Green, at the head of Schoolboy Rapids, which are about 110 feet upstream from Swallow Cave and also on the north bank. The view is downstream (west), toward the entrance to the cave. The camera station was located by means of parallax of the cliff over the cave with respect to the crags on the distant skyline. The large block of limestone against which the cottonwood log was propped in 1871 has since toppled into the water. We could make no positive correlation of other foreground boulders, although the larger ones may still be present but rotated into different positions. Small changes have occurred in the profile of the foreground cliff. As in the previous pair of pictures, the canyon wall is formed of rocks of Pennsylvanian and Mississippian age.

The light-colored crags on the skyline are Weber Sandstone, and the underlying darker cliffs and bluffs high on the canyon wall are Morgan Formation. The slopes just below the Morgan are the Doughnut Shale. The lower cliffs and slopes are Humbug Formation. A conspicuous Pleistocene fan, hidden by trees in the 1871 picture, covers part of the Humbug near the center of our photograph and is responsible for the upper end of Schoolboy Rapids. Outcrops of Madison Limestone are seen near river level.

July 28, 1968

Split Mountain 7½' quadrangle, Utah

July 10, 1871, morning

Split Mountain Canyon
Camera Station 694

This Beaman view is almost identical with that of Beaman photograph 690, but we have included it to show some interesting differences in parallax and magnification of distant detail. The foreground of 694 seems to indicate that the camera was placed closer to the rock pedestal than for 690, but the lower position of the pedestal in relation to the distant canyon wall indicates that the camera was placed farther back. Comparison of the skyline height of the near cliff on the right with the skyline height of the distant canyon wall shows the near cliff to be lower, also indicating that the position for 694 was farther back.

The rock pedestal is about the same size in both 690 and 694, but the distant canyon wall is about twice as large. This is a result of Beaman's having used a longer focal-length lens, which covers a much smaller angle of view than that of the shorter focal-length lens used to take photograph 690.

122

July 28, 1968

Split Mountain 7½' quadrangle, Utah

July 10, 1871, morning

Split Mountain Canyon
Camera Station 545

These pictures were taken from Swallow Cave near Schoolboy Rapids. The entrance to the cave, facing up-river, is shown in the two previous pairs of photographs. The cave was formed in cherty limestone beds of the upper part of the Madison Limestone. In Stephen's picture, taken in the early afternoon, the sun does not reach the cave entrance. Shoemaker's son Patrick is posing in place of the figure in Beaman's photograph, who was possibly Thompson.

The cave narrows as it follows a joint in the rock, but it extends back for an estimated 150 feet. The silhouette of the cave entrance is nearly the same today as in 1871, but some slabs of rock have loosened from its walls, and a large chunk of rock supporting Beaman's camera station has evidently fallen off. Shoemaker had to hold Stephens by the belt so that he could lean out over the water in the cave to recapture the camera position as closely as possible.

The boulder, now washed away, that is prominent in the foreground of Beaman's picture is limestone that had fallen from the entrance of the cave. The large cottonwood beyond it is now gone, but junipers are growing on the bank as they were in Beaman's day. The near bank is of Holocene alluvium. The eroded Pleistocene fan shown in the previous pairs of photographs is visible on the far (south) bank of the river. Rocks along the skyline are chiefly Morgan Formation. Underlying them are the Doughnut Shale and Humbug Formation, much of which are mantled by talus, landslide debris, and colluvium. The grassy slope on the fan is more extensively covered by piñon and juniper today than in 1871, and the pygmy forest on the more distant slopes also appears to be denser.

July 28, 1968

Split Mountain 7½' quadrangle, Utah

July 11, 1871, morning

Split Mountain Canyon
Camera Station 702

The camera station is on the north wall of lower Split Mountain Canyon, about thirteen miles upriver from the Dinosaur National Monument Campground and Boat Ramp. The view is west-northwest.

The crags on the skyline are composed of the lower two-thirds of the Weber Sandstone. The tallest crag projects about six hundred feet above the base of the Weber. The contact of the Weber on the Morgan Formation can be seen at the top of the more massive, darker wall beneath the crags. About five hundred feet of Morgan is exposed below this contact. In the right half of the pictures is Pleistocene landslide debris that extends down from a prominent ledge of Morgan into the foreground. This landslide forms a bench about one-and-a-half miles long on the north wall of the canyon. The sliding occurred on shales of Mississippian age that dip toward the river. The slide is stable today; evidently moisture from a higher precipitation rate in Pleistocene time was required to lubricate it.

July 28, 1968

Dinosaur Quarry 7½' quadrangle, Utah

127

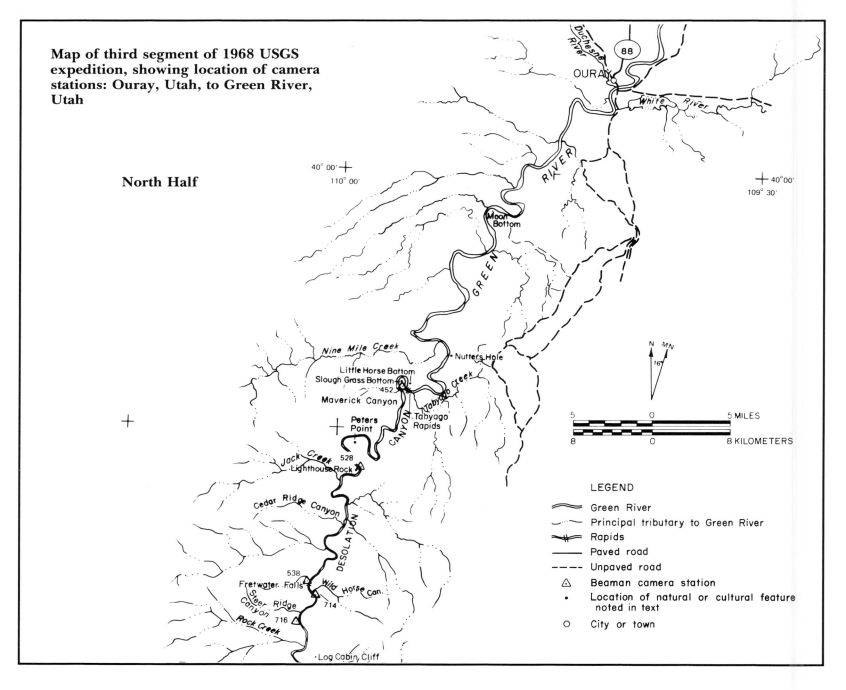

Map of third segment of 1968 USGS expedition, showing location of camera stations: Ouray, Utah, to Green River, Utah

North Half

LEGEND

Green River

Principal tributary to Green River

Rapids

Paved road

Unpaved road

△ Beaman camera station

• Location of natural or cultural feature noted in text

○ City or town

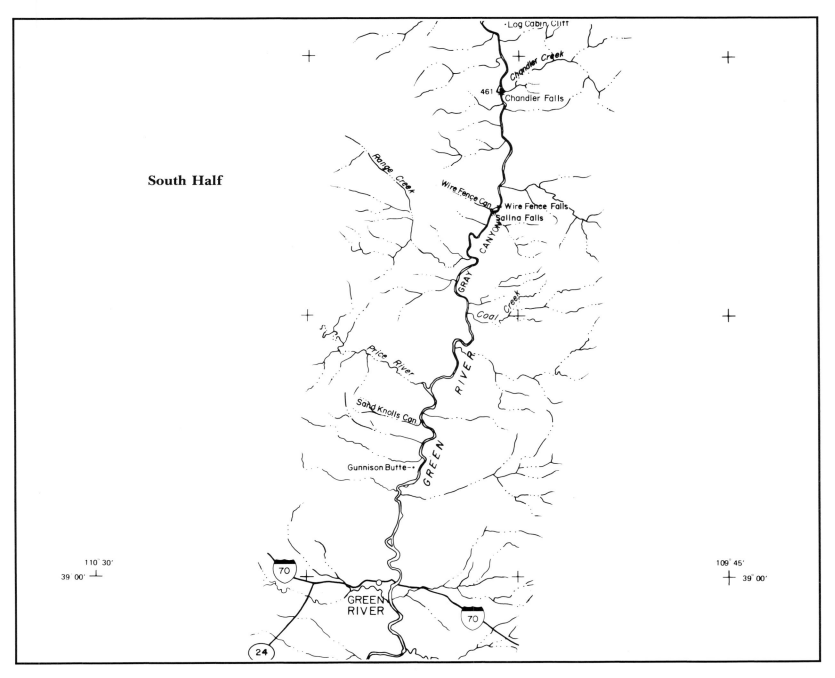

South Half

Table III: Third Segment, 1968 Expedition

Geologic formations mentioned in captions of photographs taken between Ouray, Utah, and Green River, Utah.

Geologic Era	Geologic Period	Formation or Rock Type	Remarks (Numbers are those of Beaman Photographs)
CENOZOIC	QUATERNARY	Holocene alluvium	Foreground bank in 714 and downriver at 716 and 461.
	TERTIARY	Green River Formation	Forms cliffs at Slough Grass and Little Horse bottoms in Desolation Canyon (452); also prominent in downriver pictures.
		Horse Bench Bed	Caps skyline buttes in 452.
		Wasatch Formation	Upper part forms part of Lighthouse Rock (528); also prominent downstream in Desolation Canyon.

Chapter Three

Ouray, Utah, to Green River, Utah

Having skipped over most of a section of the Green River that is sluggish, meandering, and relatively unscenic, we began the third segment at Ouray, Utah, on the edge of the Uintah and Ouray Indian Reservation.

We launched our two inflatable rafts near a bridge over Utah State Highway 88 across the river from the Ouray Trading Post. Mosquitoes swarmed over the crew members on that August day and on many of the following days.

The river meanders slowly across the wide Uintah Valley before entering Desolation Canyon. Even within the canyon, it is wide and smooth, muddy and murky, flowing along at about two miles an hour. Above the river, the sandstone, siltstone, shale, and conglomerate of the Duchesne River Formation contribute to a rather drab landscape of low relief that moved Powell to name the canyon "Desolation." Here the canyon walls are a quarter- to a half-mile apart. As we floated downriver, increasingly older rocks came into view. Below the Duchesne River Formation is the Uinta Formation, whose reddish-brown siltstone and sandstone cap many buttes and mesas in the canyon area. Below the Uinta is the Green River Formation; its light-brown sandstone forms cliffs separated by steep slopes of greenish-gray shale.

By the time we reached the first Beaman camera station (452) in Desolation Canyon, the Duchesne River and Uinta Formations had receded into the distant background; all rocks viewed from this vantage point are shale, sandstone, and marlstone of the Green River Formation. We did not encounter rapids until we had covered almost a third of this segment and reached those below Tabyago Creek.

As we dropped lower in the geologic section, past the meander of Peters Point, we reached the little pinnacle of Lighthouse Rock. Here the greenish-gray rocks of the Green River Formation give way to red rocks of the underlying Wasatch Formation. The lower part of the Green River Formation still makes up the upper canyon walls, however, and forms many pinnacles and unusual erosional remnants that produce an irregular, jagged skyline along this part of the river. Rapids become more and more common downstream.

At Fretwater Falls, a grove of cottonwood trees thirty to forty feet high with trunks at least six inches thick has grown up in front of Beaman's camera station 538. Rapids are closer together in this section of river, and some have fairly high standing waves. At Rock Creek, a short distance upstream from Powell's "Log Cabin Cliff," a spring of cold water provides refreshment for weary travelers. We saw abandoned ranch buildings some distance back from the west bank of the river; a stone house, a log cabin (blacksmith shop), and an old corral document the history of early settlements along the Green.

The stop at Chandler Falls (Beaman station 461) enabled us to view Chandler Arch, which stands across the river from the east bank, and it is a most impressive sight. Although described by Thompson in his journal for August 19, 1871, the arch is not visible from rafts or boats in midstream. To see it, we took a road that parallels the left bank of the river, rounds a bend, and heads up Chandler Creek. Here we saw evidence of past habitation in the form of a brownstone chimney, broken-down corrals, and a few scattered pieces of weathered lumber.

The river cuts still lower in the geologic section as Desolation Canyon continues into Gray Canyon. Cretaceous rocks—mostly sandstone with some shale, limestone, and a few thin coal seams—make up the drab canyon walls, which are lower than those in Desolation Canyon. Rapids, though, are still more common and more turbulent.

Gunnison Butte, a prominent landmark near the west bank of the river, marks the rendezvous where Major Powell rejoined his river expedition on August 29, 1871, after journeying to Salt Lake City for supplies. The campsite was across the river and upstream from the butte.

We ended the third segment of our trip a short distance downstream from Gunnison Butte, at a diversion dam just above the town of Green River, Utah.

Floating on a quiet stretch of the Green River in Desolation Canyon. Gene Shoemaker is at the left with Orson Anderson at the oars. Daughter Linda with the wide-brimmed hat and Gene's wife, Carolyn, make up the rest of the crew.

Desolation Canyon
Camera Station 452

August 10, 1871, morning

The camera station is on a point in the center of an incised meander above Slough Grass and Little Horse bottoms, and we are looking southwest toward the mouth and fan of Maverick Canyon. The river is flowing toward the left. The location of the station was determined from parallax between a foreground point in the lower left corner of Beaman's picture and features on the alluvial fan and canyon walls in the distance.

All of the rocks in view are shale, sandstone, and marlstone of the Green River Formation. Its upper oil-shale zone can be seen in the upper part of the canyon wall; the zone forms the second ledge above the most widespread light band in the Beaman picture. A sandstone ledge capping the buttes on the skyline in the middle of the pictures is the Horse Bench Bed of the Green River. No conspicuous changes were noted in the canyon walls, but the

August 4, 1968

distributary system on the Maverick Canyon fan has changed somewhat since 1871. Details of the larger drainage channels on the right side of the fan have been modified. A channel in the center of the fan has been abandoned entirely, and an old, indistinct drainage on the left side has been rejuvenated.

The bank of the river channel in the lower right corner of the pictures, upstream from the fan, has shifted northeast about 200 feet, but the old bank is still recognizable as a distinct line with box elders scattered along it. Just downstream, though, the river bank appears unchanged.

Beaman probably made a nearly complete panorama from this vantage point. Entries in the Powell expedition journals, chiefly Thompson's, suggest that Beaman made nine successful plates, of which only one broken plate remains in the Hillers Collection.

Nutters Hole 15' quadrangle, Utah

135

August 11, 1871, morning

Desolation Canyon
Camera Station 528

These pictures of Lighthouse Rock were taken from the northeast bank of the river at the site of Powell camp no. 36, below the big meander of Peters Point. The view is downstream, slightly east of south.

The river banks are slightly steeper today but are still lined with box elders, whose limbs now hang farther down into the scene. We had to break away a few in order to obtain Stephens's matching picture. A few young tamarisks are also present now.

Rocks in view include the uppermost part of the Wasatch Formation and roughly the lower half of the overlying Green River Formation. Here the contact between the two formations is arbitrary because the rocks are similar: both are relatively massive sandstone beds and minor interbeds of mudstone, both deposited by rivers. The contact is placed at the change in color of the mudstone: that of the Wasatch, deposited in a well-drained, oxygen-rich environment, is red. Mudstone of the Green River Formation, laid down under oxygen-poor conditions, is greenish gray. The color change occurs at the foot of the uppermost cliff under the pinnacle of Lighthouse Rock.

The higher beds in the distant canyon wall are lake deposits of mudstone, marlstone, shale, and minor sandstone, which are more typical of the Green

River Formation. The ledge on the skyline is formed of sandstone just above the upper oil-shale zone noted at camera station 452.

August 5, 1968

Firewater Canyon 15' quadrangle, Utah

137

August 14, 1871, early afternoon

Desolation Canyon
Camera Station 538

We are here on the west bank of the river at Fretwater Falls, about a mile above the mouth of Wild Horse Canyon. We based our camera location on parallax between features on the high cliff close to the river and the more distant canyon wall. View is to the northeast across the main part of the falls.

The foreground is the lower part of a dissected fan about fifteen feet above the river. Since 1871, sand blown by wind from a downstream bar has covered some of the foreground boulders, and a grove of cottonwoods has grown up. A small group of cottonwoods on the far bank is still present as are most of the junipers on the slope, now more numerous.

All rocks in view are Wasatch Formation. Those on the skyline on the left are chiefly sandstone in the uppermost part of the Wasatch. The slope just above the river is composed of a thick red mudstone sequence.

138

August 6, 1968

Flat Canyon 15′ quadrangle, Utah

August 15, 1871, morning

Desolation Canyon
Camera Station 714

The camera station is on the east bank of the Green just below the mouth of Wild Horse Canyon; the view is downstream, slightly east of south. Our camera station was located by parallax between promontories in the distant view.

Comparison of the pictures shows a few changes. A small gully, visible about thirty feet in front of Beaman's camera station, is less well defined today. The near river bank has been eroded back ten to possibly twenty feet in places, but the downstream end of the bar in the river seems to be in almost the same position. A snaggy old cottonwood in the middle distance is still present, although a limb seems to have broken off. The old fellow probably marks the site of Powell camp no. 39.

The foreground bank is Holocene alluvium. Its upper part is composed of sediments derived chiefly from the canyon walls and the side canyons. The canyon walls in the middle distance are interbedded sandstone and mudstone of the Wasatch Formation. Shale and sandstone of the Green River Formation make up the distant skyline ridges and the top third of the slopes below them.

140

August 6, 1968

Flat Canyon 15' quadrangle, Utah

Desolation Canyon
Camera Station 716

These pictures were taken on the west bank of the river about half a mile below the mouth of Steer Ridge Canyon. The view is upriver, slightly east of north, across the head of a rapid. Beaman took his picture just above Powell's camp no. 40. The station is on the edge of a dissected Holocene fan, which is responsible for the rapid and is the source of the large boulders—one twenty feet across—shown in the pictures. A boulder close to the water's edge appears to have been upended since 1871.

We located the camera station by parallax between various parts of the canyon wall, because we could make no positive identification of any of the foreground objects. Although the water stage in Beaman's photograph appears to be lower than in ours, no water was flowing on the near side of the boulder bar when we were there. The gradient of the main part of the rapid, therefore, may have been lowered since 1871, or the amount of foreground gravel may have been augmented by a flash flood from one of the side canyons, which would also account for the upending of the large boulder. The canyon walls are all Wasatch Formation. Sandstone beds topping the distance butte are near the top of the Wasatch. A mudstone slope low on this butte corresponds to the slope noted earlier in photograph 538.

August 17, 1871, morning

142

August 6, 1968

Flat Canyon 15' quadrangle, Utah

Desolation Canyon
Camera Station 461

August 19, 1871, early afternoon

This camera station is near the foot of Chandler Falls on the west bank of the Green, just below the mouth of Chandler Creek. The view is to the northeast. Beaman's picture shows the boats being lined down the falls; the figure in the nearest boat is Steward. A juniper against the skyline at the left edge of Stephens's picture was also in Beaman's photograph (before cropping); now it clings precariously to the rim of the bank, part of which has slumped away since 1871.

The mass of boulders along the bank has almost completely changed; most of them have been swept downstream as the river undercut the bank. We recognized only one boulder—the large one near the head of the falls that forms a point around which the river swings. The exact camera station could

August 7, 1968

not be occupied because it had been cut away by the river, but Stephens's photograph was probably taken within two feet of the correct position, confirmed by the juniper noted above.

The water stage was somewhat lower at the time of Beaman's photograph than at the time of ours. Across the river, on the right side of our picture, young cottonwoods have advanced well out onto the gravel bank. The trees on the far canyon wall are junipers.

The canyon walls are sandstone and mudstone of the Wasatch Formation. Two fairly thick sequences of red mudstone and minor interbedded sandstone form slopes low on the walls. The bank on the left side is an alluvial terrace, apparently Holocene, formed as a fan at the mouth of a small side canyon.

Range Creek 15′ quadrangle, Utah

145

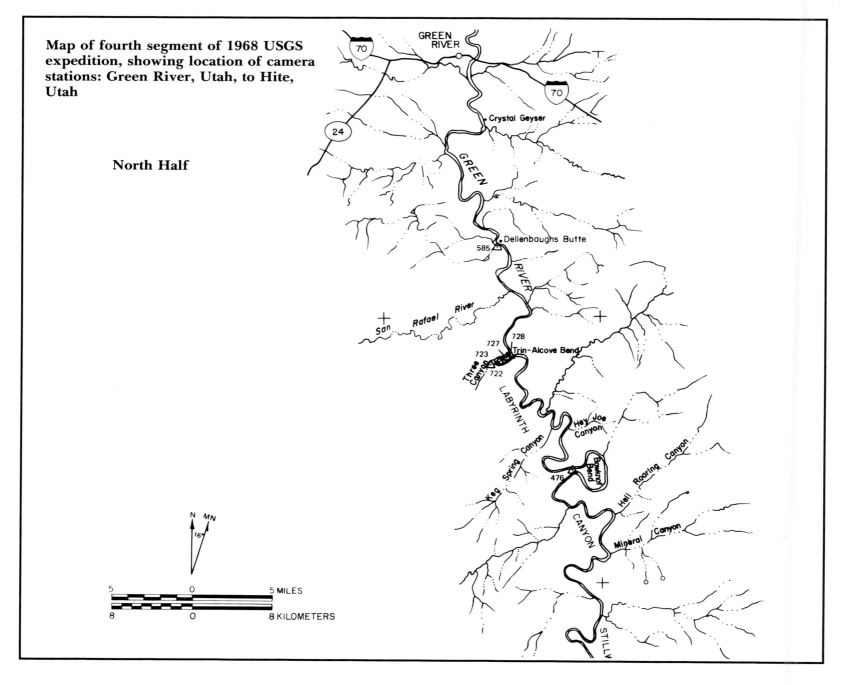

Map of fourth segment of 1968 USGS expedition, showing location of camera stations: Green River, Utah, to Hite, Utah

North Half

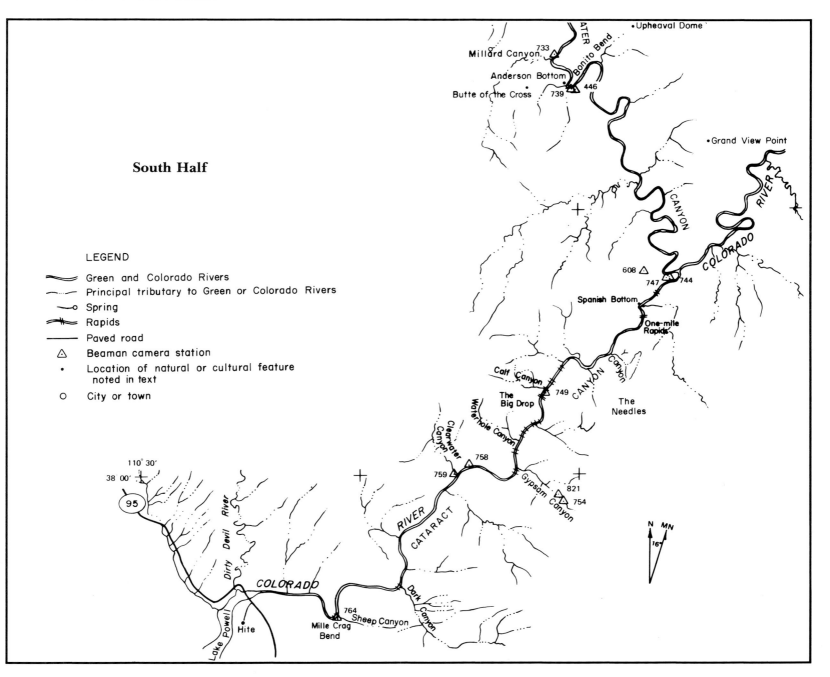

South Half

LEGEND

〜〜〜 Green and Colorado Rivers
〜· 〜· Principal tributary to Green or Colorado Rivers
〜─o Spring
╫╫╫ Rapids
─── Paved road
△ Beaman camera station
· Location of natural or cultural feature
   noted in text
O City or town

Table IV: Fourth Segment, 1968 Expedition

Geologic formations mentioned in captions of photographs taken between Green River, Utah, and Hite, Utah.

Geologic Era	Geologic Period	Formation or Rock Type	Remarks (Numbers are those of Beaman Photographs)
CENOZOIC	QUATERNARY	Holocene alluvium	In 723 and at Anderson Bottom (739).
		Pleistocene terrace gravel	Caps cliff above Green River in Labyrinth Canyon (727, 728).
		Pleistocene talus	In Cataract Canyon (749) and Gypsum Canyon (754, 821).
MESOZOIC	JURASSIC	Morrison Formation Salt Wash Member	Basal beds form uppermost part of Dellenbaughs Butte (585).
		Summerville Formation	Forms upper cliff of Dellenbaughs Butte (585).
		Curtis Formation	Forms slope and lower cliff of Dellenbaughs Butte (585).
		Entrada Sandstone	Uppermost 15 feet exposed at base of Dellenbaughs Butte (585); at Trin-Alcove Bend (727, 728).
		Carmel Formation	On skyline in Labyrinth Canyon (727, 728).
	JURASSIC AND TRIASSIC	Navajo Sandstone	Makes up all rocks in view in 722 and 723 at Trin-Alcove Bend; seen downriver as far as mouth of Green (744).
	TRIASSIC	Kayenta Formation	Lower beds cap cliffs at Bowknot Bend (476) and upper part of Butte of the Cross (733); seen downriver as far as mouth of Green (744).

Geologic Era	Geologic Period	Formation or Rock Type	Remarks (Numbers are those of Beaman Photographs)
MESOZOIC	TRIASSIC	Wingate Sandstone	Large cliff at Bowknot Bend (476); forms most of Butte of the Cross (733); seen downriver as far as mouth of Green (744).
		Chinle Formation	In Bonito Bend area (733, 739, 446) and downriver as far as mouth of Green (744).
		so-called Black Ledge	Below Wingate cliff at Bowknot Bend (476).
		Moenkopi Formation	Lowermost part of section at Bowknot Bend (476); upper members form lower part of Butte of the Cross (733); in Bonito Bend area (466); also near mouth of Green (744).
		Hoskinnini Member	Ledge-forming; in Bonito Bend area (733, 739).
PALEOZOIC	PERMIAN	Cutler Formation White Rim Sandstone Member	Well-displayed at Bonito Bend (733, 739, 446) and near mouth of Green (744).
		Organ Rock Member	At Bonito Bend (739, 446) and near mouth of Green (446).
		Cedar Mesa Sandstone Member and equivalent	Forms rim of lower Stillwater Canyon (744); makes up all rocks in view in 747 and 608; seen in all photographs from 749 on; forms Mille Crags (764).
		Halgaito Member	In Cataract Canyon (758, 759).
	PERMIAN AND PENNSYLVANIAN	Rico Formation	Forms slopes and ledges in Stillwater Canyon (744), Gypsum Canyon (754, 821), and Cataract Canyon (749, 758, 759, 764).
	PENNSYLVANIAN	Hermosa Formation	Lowermost cliff in lower Stillwater Canyon (744); in all photographs from 749 on.
		Paradox Member	In Cataract Canyon (749) and Gypsum Canyon (754, 821).

D. Julien's inscription of May 3, 1836, in Hell Roaring Canyon, a side canyon off the left bank (facing downstream) of Stillwater Canyon of the Green River. Felix Mutschler (left) and Gene Shoemaker. Julien, a trapper, was one of the first men to explore the Green and Colorado rivers—and leave his mark.

150

Chapter Four

Green River, Utah, to Hite, Utah

Our departure point for this segment was the Green River State Park at Green River, Utah. Our new boat crew assembled in the beautiful, well-kept campground situated in a grove of mature cottonwood trees.

A fair current sweeps past the boat ramp, but the surface of the water is smooth. Willow and tamarisk grow thickly along the shoreline. Looking upstream, we could see Gunnison Butte, near the mouth of Gray Canyon.

About two miles downstream, the west bank has low relief. The tan, ledge-forming Salt Wash Member of the Jurassic Morrison Formation forms the lower sixty feet of the east canyon wall. Just ahead, the overlying banded red and gray Brushy Basin Shale Member of the Morrison forms slopes about 200 feet high above the Salt Wash and is in turn capped by the brown Burro Canyon Formation. However, the rocks dip gently upstream, and as we floated downriver we saw progressively older rocks.

About four-and-a-half miles downstream from the Green River boat ramp is Crystal Geyser, which erupts water from a hole drilled for petroleum exploration. Eruptions are periodic, caused by buildup of carbon-dioxide pressure. Red, brown, and yellow terraces of calcareous tufa deposited by the well flow surround the site.

Dellenbaughs Butte, named for the youngest member of the Powell expedition, is a major landmark on the east bank about twenty-eight miles downstream from the town of Green River (Beaman photograph 585). The cliff and the slopes and ledges beneath it expose a thick Jurassic section: the Entrada Sandstone at the base, then the Curtis and Summerville Formations, and Salt Wash at the top.

Two miles below the mouth of the San Rafael River, exposures of Navajo Sandstone appear below the Entrada and form cliffs for about a mile before Trin-Alcove is reached.

Trin-Alcove is another spectacular landmark named by Powell: three steep-walled canyons, entirely in Navajo Sandstone, converge on the outside of Trin-Alcove Bend. The Powell party camped under a large cottonwood tree at the confluence of these three canyons (Beaman camera station 723) in early September of 1871. Across the river, Beaman took another picture (722) of the Trin-Alcove. The 1871 party explored the high ridge in the center of Trin-Alcove Bend, taking additional pictures (727 and 728) from its crest. We found Indian arrow points in drainage catch basins on this sandstone ridge.

As we floated downstream for two or three hours after leaving Trin-Alcove Bend, we could see older and older rocks. The base of the Navajo was exposed, then the purplish-red slabby sandstone of the

Kayenta Formation, and then the salmon-colored, cliff-forming Wingate Sandstone. Willow and tamarisk still dominated the vegetation of the river banks and islands.

We found modern petroglyphs on a vertical Wingate face about one-and-a-half miles below Keg Spring Canyon, about one-and-three-quarters miles above Hey Joe Canyon. Among other inscriptions on the smooth sandstone surface visible from the river is "Nevills Expedition, 6-21-28."

Bowknot Bend, an incised meander easily recognized by its distinctive shape, was examined by the Powell party. Beaman climbed to the summit of the low ridge at the "knot" for an east-facing picture (476) of Wingate cliffs. Behind his camera station, under a protecting ledge of sandstone, we found many inscriptions of early river runners, including "Kolb Bros., Oct. 1911." To these records of years gone by has been added "The Beaman and Hillers Photograph Album Expedition, August 19, 1968," carved by the then twelve-year-old Patrick Shoemaker, the son of E. M. Shoemaker.

River runners who pause in their journey to go a few hundred yards up Hell Roaring Canyon, farther downstream, can view what may be the oldest inscription along the river, carved by a white trapper-explorer who roamed the Green River drainage. On the south side of the canyon, about fifty feet above the valley floor, is the notation

3 mai 1836 D. Julien

Downstream, the canyon widens and the buttes and mesas are much farther back from the river, appearing as solitary structures on the skyline. Reddish-brown siltstone and sandstone of the Moenkopi Formation are exposed. Indian ruins—two circular towers on a rounded promontory of grayish-green and tan Moenkopi rocks—can be seen on the east bank about 200 feet above the river.

Massive bluffs sixty to seventy feet high of the White Rim Sandstone Member, the uppermost unit of the Cutler Formation, border the river where we stopped to view the magnificent Butte of the Cross. A climb up the west bank and up the rim of a sandstone bluff brought us onto a slightly undulatory, eroded surface of the White Rim. The Butte of the Cross is actually composed of two narrow, bladelike buttes. When seen in line, one butte viewed "edge on" forms the upright post, while the more distant butte seen "side on" makes up the arms. Wingate Sandstone capped by Kayenta Formation forms the top of the cross as well as the arms. The viewpoint and camera station are about half a mile above Millard Canyon.

Anderson Bottom, which adjoins Bonito Bend, was recognized as a cutoff meander by the Powell party when they stopped here on September 12. The cliffs both inside and outside Bonito Bend (photograph 446) are formed of the White Rim Sandstone Member. Underlying cliffs and steep slopes are the Organ Rock Member of the Cutler. Just to the left of the area shown in photograph 446 is the abandoned Anderson ranch, where we found fresh spring water.

One mile above the confluence of the Green and Colorado rivers, a steep canyon on the west provides access to the canyon rim 1,200 feet above the river. Here Beaman took photograph 608. We rated the view from the rim as among the most spectacular on the Colorado Plateau. To the southwest is the "Rock Forest" or the "Land of Standing Rocks," a maze of pinnacles, towers, mesas, and canyons (Beaman photograph 747). The distant view to the southeast shows the spires, pinnacles, and fins of the Needles. All of these strangely eroded features are formed in the red and buff sandstone of the Cedar Mesa Sandstone Member of the Cutler Formation.

The oldest rocks exposed in cliffs along the Green River here are part of the Hermosa Formation, overlain by slopes and ledges of the Rico Formation. Basal beds of Cutler form the canyon rims in the lower reaches of the Green.

Spanish Bottom, just below the confluence of the Green and Colorado rivers, is the logical place to camp and rest before running the thundering rapids of Cataract Canyon. Below the confluence, the Colorado flows along the axis of the Meander Anticline, a major structural feature. Rocks of the Hermosa and Rico Formations are highly fractured along this fold. Above Spanish Bottom, the skyline is made ragged by irregular pinnacles carved from the Cedar Mesa Sandstone Member. We could see these pinnacles from various points on the river.

Rapids in Cataract Canyon are more impressive than most encountered upstream. After getting through about twenty fair-sized rapids in Cataract, we reached "The Big Drop," about half a mile below Calf Canyon. Without question, it is the most notorious rapid in this part of the canyon. The crew of one of our two boats was thrown out when the boat plunged into a large hole at the base of the drop. The boat did not overturn, but the motion was so abrupt that no one remained aboard.

Gypsum Canyon, coming in from the east farther downstream, was explored for four miles or more by the Powell party. Our photographs at camera stations 754 and 821, about three-and-a-half miles up Gypsum Canyon, record the effects of massive rock falls from the canyon walls.

Gypsum and gray shale of the Paradox Member of the Hermosa Formation are exposed from the vicinity of Calf Canyon Rapids (Beaman photograph 749) on downstream to below Gypsum Canyon. Walls of Cataract Canyon tower nearly two thousand feet above the river in the vicinity of Beaman camera station 758. Mille Crags, on Mille Crag Bend, are striking features that form a jagged skyline of sandstone crags cut from the Cedar Mesa Sandstone Member of the Cutler. Beaman photograph 764 shows the crags from the opposite shore of the river. This station is now flooded by the waters of Lake Powell.

The fourth segment of our trip ended just uplake from Hite Marina. On the opposite shore are well-preserved prehistoric Indian ruins, described in the journals of the Powell expedition. On a sandstone face nearby, an inscription was made by a member of the Powell party on June 27, 1872. (This locality was revisited at that time by some of the Powell expedition members when, after completing the river trip, they traveled overland from Kanab to retrieve the *Cañonita*, which had been cached at the mouth of the Dirty Devil River nearby.)

Green River
Camera Station 585

These pictures show Dellenbaughs Butte, on the east bank of the Green about four miles above the mouth of the San Rafael River. The location of our station was based on parallax among various features on the butte and is probably accurate to within about five feet.

The butte displays about two hundred feet of the geologic section. Beginning just above river level, the uppermost fifteen feet of the Entrada Sandstone forms the low cliffs seen in gaps between the bushes. The overlying cliff and the slope above it are gray sandstone and shale of the Curtis Formation. In the higher cliff, we see almost 150 feet of reddish-brown to brown shale, mudstone, and sandstone of the Summerville Formation, which is close to its maximum thickness here. The Salt Wash Member of the Morrison Formation forms the uppermost slope and the small cap at the very top, part of which appears to have broken off since 1871.

Just above the river bank are low, partly stabilized modern dunes. Those on the right have been eroded by the river since Powell's expedition, but those on the left appear to be about the same today.

September 4, 1871, early afternoon

154

August 17, 1968

Green River 15' quadrangle, Utah

Labyrinth Canyon
Camera Station 722

September 8, 1871, morning

The camera station is on the east bank of the Green River opposite the mouth of Trin-Alcove (Three Canyons). The view is across the river, looking up the Trin-Alcove. We located the camera station by parallax among the alcoves and our position is probably accurate to about five feet, although offset slightly toward the water's edge because a heavy thicket of willows obscures the view from Beaman's station higher on the bank. On the opposite bank, Beaman's willows have largely been replaced by tamarisks. The mouth of Trin-Alcove Creek has shifted about fifty feet upstream since 1871.

All the rock in view is Navajo Sandstone, representing almost the entire thickness (about four hundred feet) of the formation. In its lower part here are a few thin purplish (Kayenta-like) sandstone interbeds, one of which, on the west bank above Trin-Alcove, contains molds and casts of gypsum crystals. We detected no changes in the bedrock.

August 18, 1968

Bowknot Bend 15' quadrangle, Utah

September 8, 1871, afternoon

Labyrinth Canyon
Camera Station 723

Following Beaman's route, we have now crossed the river and are on its southwest bank, looking west-northwest up the Trin-Alcove. Our exact location is based on the distinctive foreground rock, lower right, and an abutment of the canyon wall in the lower left.

The vegetation has changed less than one would expect. An oak tree in the foreground of Beaman's picture is now dead, but young shoots have sprung up at its base. Most of the trees on the floor of the canyon are cottonwoods, and at least two of the larger ones in the 1871 picture are still alive.

The canyon is floored by alluvium derived from the Navajo Sandstone, now dissected by a gully ten feet deep. No dissection, however, can be seen in the Beaman photograph. All the visible bedrock is Navajo Sandstone. The stain (desert varnish) on the right canyon wall appears to be about the same today, although some stained rock may have spalled away. A few thin layers have spalled from the face of the rock in the right foreground. Note the ripple marks on this rock visible in Stephens's picture; they indicate the action of shallow water on loose sandy sediment, as on beaches.

August 17, 1968

Bowknot Bend 15' quadrangle, Utah

September 8, 1871, morning

Labyrinth Canyon
Camera Station 727

We are here on a ridge in the center of Trin-Alcove Bend, looking slightly south of west toward the outside of the bend. The Trin-Alcove can be seen just left of the center of the pictures. The location of Beaman's camera station was based on the rounded bench of Navajo Sandstone in the foreground. The shrub on the right is still alive, but the split boulder behind it is gone, and no obvious remains of it were found downslope.

In the middle distance is a wind-eroded, sand-filled hollow in which we found several arrowheads and crude handtools that may represent more than one culture. On the sandstone bench, we also found pieces of wood that may have been brought in by early Indians.

Most of the rocks in the field of view, including those of the far canyon walls, are Navajo Sandstone. The Carmel Formation forms a dark strip along the skyline on the right side of the pictures, and the Entrada Sandstone forms the dark, low buttes on the skyline in the center, about two-and-a-half miles distant. The dark cap on the ridge point just in front of us is Pleistocene terrace gravel more than 280 feet above the river.

160

August 18, 1968

Bowknot Bend 15' quadrangle, Utah

September 8, 1871, morning

Labyrinth Canyon
Camera Station 728

This camera station is also on the south bank, on the same ridge as the last but about a mile upstream from Trin-Alcove Bend. The view is upstream to the north-northwest. Location of the camera station was based on sandstone points in the foreground, but Stephens's camera was offset slightly along the line of sight in order to show the rim of the cliff in proper perspective. Delicate features on the cliff rim appear to be virtually unchanged.

We do see significant changes along the west (left) bank of the river. The present bank extends much farther out into the river over what was once an island, and the large bars are mostly gone. The east bank may have been eroded back slightly near the right edge of the picture.

The canyon walls are cut entirely in Navajo Sandstone. High-level terrace gravel forms a distinct dark mound just back from the rim of the right canyon wall, and it also occurs on a low, dark ledge just back from the edge of the left wall.

On the skyline on the left is a series of Entrada Sandstone cliffs and bluffs. The Entrada and the Carmel Formation can also be seen just above the Navajo near the center of our picture. To the right, the distant Beckwith Plateau and Roan Cliffs and the eastern

Book Cliffs also stand out clearly on the skyline in the 1968 picture but were apparently obscured by haze in 1871.

August 18, 1968

Bowknot Bend 15' quadrangle, Utah

163

Labyrinth Canyon
Camera Station 476

September 10, 1871, morning

The camera station is at the foot of a Wingate Sandstone cliff at the west end of the ridge forming the "knot" of Bowknot Bend. We are looking east, toward the "bow"; the upstream segment of the river is on the left.

Bedrock exposed in the canyon walls ranges from the upper part of the Moenkopi Formation up to the lower part of the Kayenta Formation. The large cliff is Wingate Sandstone capped by the slabby Kayenta. All the foreground and most of the canyon walls below the cliff are Chinle Formation, whose "Black Ledge" is prominent below the Wingate.

Moenkopi is exposed between talus and slides on the lower canyon walls. No changes were observed in the bedrock.

Vegetation along the river, however, has changed considerably. On the river bottom at the left, a dense thicket of tamarisk, serviceberry, and oak has filled in the open area behind the willows. On the right, an island has shifted downstream and is now overgrown with tamarisk and fringed with willow. The river was much lower at the time of Beaman's photograph, and considerably more sand banks and bars were exposed than when we were there.

164

August 19, 1968

Bowknot Bend 15' quadrangle, Utah

September 12, 1871, about noon

Stillwater Canyon
Camera Station 733

These pictures were taken on the west bank of the Green about half a mile above Millard Canyon. Exact location of the station was based on features in the foreground ledge. The view is to the southwest, toward the Butte of the Cross.

The foreground bench and other light-colored, low-lying rocks are the White Rim Sandstone Member of the Cutler Formation. In the middle distance beyond the river bend, the first low ledge above the White Rim is the Hoskinnini Member of the Moenkopi Formation, surmounted by a series of low bluffs and spurs of the Moenkopi's middle ledge-forming member. The broad base of the Butte of the Cross is formed by the upper slope-forming member of the Moenkopi, here bleached to a light tan. An upper ledge-forming unit of the Moenkopi caps the light-colored slopes. The next ledge up is the basal sandstone of the Chinle Formation. Wingate Sandstone capped by the Kayenta Formation forms the arms and the upright of the cross.

Other cliffs and buttes of Wingate capped by Kayenta can be seen behind the cross to its right and left. Navajo Sandstone forms the skyline on the extreme right of the pictures and also a small ear-shaped point just to the left of the Butte of the Cross. Detailed features of the foreground ledge have

166

scarcely changed since 1871. Pockets of sand provide footing for low shrubs that are in nearly the same positions today. The river bottom in the center of the pictures has almost the same vegetation pattern: willows along the bank and several species of brush cover most of the bottom, and serviceberries and other trees grow higher up. The chief difference between the pictures is the invasion of tamarisk.

August 21, 1968

The Spur 15' quadrangle, Utah

167

Stillwater Canyon

Camera Station 739

September 13, 1871, morning

This camera station is on the south bank of the Green at Bonito Bend. The view is southwest, directly away from the river.

Foreground blocks are eroded talus of the White Rim Sandstone Member of the Cutler Formation resting on the Organ Rock Member of the Cutler. The contact of the White Rim bedrock on the Organ Rock is in the notch at the base of the overhanging cliff on the left side of the pictures.

The White Rim cliff rising above an Organ Rock slope that is partly covered by talus forms the rim of the incised oxbow (a cutoff meander) that surrounds Anderson Bottom. This oxbow is one of the farthest downstream on the Green and Colorado rivers and was probably cut off in late Pleistocene time; Thompson, in 1871, was especially impressed with the cutoff having occurred so recently. The broad floor of the bottom near the river is covered by Holocene alluvium. Some of the small cottonwoods growing on this alluvium in 1871 are now large trees.

Buttes in the distance are formed by Wingate Sandstone capped by the Kayenta Formation. Beneath the Wingate, a complete succession of Lower to Upper Triassic strata is exposed. Of the units visible in these pictures, the lowermost is the Hoskinnini Member of the Moenkopi, which forms the inconspicuous first ledge above the White Rim on the right side of the pictures. The middle ledge-forming member of the Moenkopi caps bluffs above the Hoskinnini. A ledge at the base of the Chinle is barely perceptible on the slopes of Merrimac Butte (just right of center).

August 21, 1968

The Spur 15' quadrangle, Utah

169

Stillwater Canyon
Camera Station 446

September 13, 1871, morning

Just downstream from the previous camera station, we are still on the south bank of the river at Bonito Bend, looking northeast. The river flows to the right. The exact location of our camera station was based on rocks on the foreground slope.

Rocks in view range from the Cutler Formation up to the Navajo Sandstone. The cliffs both inside and outside the river bend are the White Rim Sandstone Member of the Cutler. The Organ Rock Member of the Cutler forms part of a steep slope just beneath the White Rim cliff. The Moenkopi and Chinle Formations form a series of benches above the river on the right limb of Bonito Bend. The Wingate Sandstone capped by Kayenta Forma-

August 21, 1968

tion forms the prominent cliff seen along most of the skyline. Above the river on the downstream part of the bend, however, the skyline is formed by a great mound of Navajo Sandstone on the edge of Upheaval Dome.

A fallen slab of the White Rim forms an ear projecting over the water at the end of the point on the inside of Bonito Bend. This slab is a remnant of the sandstone septum that separated the upstream and downstream branches of the river when it flowed around the Anderson Bottom. Just to the left of the scene is Anderson's ranch (abandoned), a source of fresh, clear spring water.

The Spur 15' Quadrangle, Utah

171

Stillwater Canyon
Camera Station 744

September 16, 1871, morning

Our last picture of the Green River was taken from the south rim of Stillwater Canyon about a mile southwest of its junction with the Colorado River. The view is north-northwest, up the Green. We located Beaman's exact camera station by parallax between the near and far rims of the canyon and by identifying rocks and vegetation—still the same—on the canyon rim just beneath the camera. Below, tamarisks have grown up along the river bank where sandbars show in Beaman's photograph.

The rocks in view range from the Hermosa Formation to the Navajo Sandstone. Along the river, the lowermost cliff is Hermosa; the top of the highest cliff is formed by the basal part of the Cedar Mesa Sandstone Member of the Cutler Formation, which here is buff sandstone with sparse interbeds of gray marine limestone. Between the lowest and highest cliffs are slopes and ledges of the Rico Formation.

At the skyline on the right is Grand View Point, a cliff of Wingate Sandstone capped by Kayenta Formation. Slopes and ledges below the Wingate are Chinle, Moenkopi, and the upper members of the Cutler. The spire on the left is Wingate capped by Kayenta. Navajo Sandstone forms isolated knobs on the skyline in the center of the pictures.

172

August 23, 1968

The Needles 15' Quadrangle, Utah

September 16, 1871, about noon

Cataract Canyon
Camera Station 747

We are now about three hundred feet southwest of the previous camera station, but on the north rim of Cataract Canyon on the opposite side of the point between the Green and Colorado Rivers. The view is to the west toward Powell's "Rock Forest." The location of our camera station was based on details of the foreground ledge, which are unchanged except for a new spall that has formed close to the camera. Note the juniper on the left in Beaman's photograph; its carcass is still in place.

The rocks in view are all part of the Cedar Mesa Sandstone Member of the Cutler Formation. The lower, light-colored sandstone is characterized by large-scale crossbedding, probably indicative of ancient sand dunes. The pinnacles are formed in red- and white-banded, parallel-bedded sandstone with minor red mudstone interbeds. A few miles to the south, these rocks grade into white sandstone more typical of the Cedar Mesa.

August 23, 1968

The Needles 15' quadrangle, Utah

September 17, 1871, morning

Stillwater Canyon
Camera Station 608

This camera station is on the edge of The Maze, named by Powell, about one-and-a-half airline miles west of the Green and Colorado rivers. We reached the station by the same route that Beaman took, up a steep side canyon on the west bank of the Green. View is to the west. The exact location of the camera station was based on features on the foreground outcrop, which are nearly the same. The trees in the middle distance that are still alive are piñon and juniper.

All the rocks in view are in the upper part of the Cedar Mesa Sandstone Member of the Cutler Formation. They are strikingly banded red and buff. Except for few thin mudstone strata, all the beds are sandstone. The foreground sandstone is red, medium to coarse grained, and crossbedded on a fairly large scale. The crossbeds probably reflect its origin as ancient sand dunes.

The scene shows several changes, as well as some features that, surprisingly, have changed very little. The water level is about the same, as is a cluster of rocks in the foreground. Just a few feet to the right, however, a seven-foot block of sandstone has tumbled down to obscure part of the immediate foreground. In the middle distance, the large blocks along the bank are mostly in place. The trees still growing along the normal highwater line on the right are hackberries. Tamarisks are now scattered along the edges of the sandbar across the river.

Rocks exposed in the canyon walls range from the upper part of the Paradox Member of the Hermosa Formation to the Cedar Mesa Sandstone Member of the Cutler Formation. Tilted beds of gypsum and gray shale of the Paradox can be seen above the hackberry trees at the right, but they are partly obscured by dissected talus. Both this talus and that on the opposite side of the river are of Pleistocene age. Just above the talus and the tilted beds of Paradox are bluffs and promontories of interbedded gray limestone, sandstone, and shale of the upper member of the Hermosa. Overlying the Hermosa, midway up the canyon walls, the Rico Formation forms a relatively uniform band of interbedded red arkosic limestone and gray limestone. Cedar Mesa forms the skyline cliffs on both sides of the canyon.

August 23, 1968

The Needles 15′ Quadrangle, Utah

Cataract Canyon
Camera Station 749

In these pictures, our first view of the mighty Colorado, the river looks deceptively peaceful, especially in Beaman's photograph. Cataract Canyon, however, lives up to its name, as described in the introduction to this segment. We are on the east bank of the river looking upstream toward Calf Canyon Rapids, just above Powell's midday dinner camp. The exact location of the camera station was based on foreground rocks.

September 22, 1871, about noon

August 25, 1968

Orange Cliffs 15' quadrangle, Utah

Gypsum Canyon
Camera Station 754

This camera station is in Gypsum Canyon about three-and-a-half miles above its mouth and about five hundred feet downstream from the first major waterfall. The view is to the north, down the canyon. The exact location of the camera station was based on foreground rocks on the talus slope.

Dramatic changes have occurred here since Beaman's day. In the middle distance are large rock falls from both sides of the canyon. A great slab has fallen away on the left, producing a deep notch in the skyline. On the right, the source of the fallen rocks is less discernible, but a huge pile of broken rock lies at the foot of the canyon wall. A sand deposit probably more than fifteen feet thick has formed in the ponded area behind the rock falls that partly dam the canyon. Upstream, however, the canyon floor seems to have been partly scoured. In the foreground, several large blocks on the talus have slid or rotated. Many of the changes may have been caused by flash floods that locally undercut the talus and canyon walls.

The rocks in view are mostly Hermosa Formation that has been folded and faulted. The high cliff, which is limestone, represents virtually all of the upper member. Below it, the ledgy slope of alternating beds of black shale and limestone is in the upper part of

September 25, 1871, near noon

180

the Paradox Member; its uppermost
bed of gypsum crops out behind the
camera station.

At the skyline, against the canyon
wall that is in shadow in the Beaman
picture, we can see the Rico Formation
and the lower part of the Cedar Mesa
Sandstone Member of the Cutler Formation.

August 26, 1968

Mouth of Dark Canyon 15′ quadrangle, Utah

181

September 25, 1871, afternoon

Gypsum Canyon
Camera Station 821

Still in Gypsum Canyon, we are here about two thousand feet downstream (north) from the previous camera station. The view is to the northwest. We located the station by parallax between features on the canyon walls in the middle distance, probably to within five feet.

All of the foreground has changed drastically as the result of the rock falls described for photograph 754. Just in front of our camera station is the deposit of sand formed as a consequence of the rock falls damming the canyon.

The uppermost black shale of the Paradox Member of the Hermosa Formation forms a short slope that is sunlit in Beaman's picture. Most of the visible canyon walls are of the upper member of the Hermosa. In the distance a point of the Cedar Mesa Sandstone Member (Cutler) and part of the underlying Rico Formation is framed between the canyon walls. A thick deposit of red Pleistocene talus has formed at the foot of the high cliff in the left center of both pictures. Such deposits are found from this point all the way down Gypsum Canyon.

August 26, 1968

Mouth of Dark Canyon 15' quadrangle, Utah

September 26, 1871, about noon

Cataract Canyon
Camera Station 758

We are back on the river now, on the north bank about a mile above Clearwater Canyon, looking upstream. Our exact camera station was based on the foreground rocks, most of which we could identify. Rocks at the extreme left, however, have probably either rotated or tumbled into the river, and some are partly obscured by tamarisks. The large driftwood log in our picture was a good twenty feet above water level, indicating a floodstage of at least this height since 1871. Such a high water level must be rare, because good-sized tamarisks are solidly rooted about five feet lower on the river bank.

The rocks in the canyon walls range from the lower part of the upper member of the Hermosa Formation to the Cedar Mesa Sandstone Member of the Cutler Formation. Cedar Mesa forms the prominent points and buttresses along the nearly vertical upper canyon wall in the distance. A short slope beneath this cliff is formed by the Halgaito Member of the Cutler. The underlying Rico Formation forms ledges and slopes on which piñon and juniper are growing. The main part of the canyon wall is nearly vertical and almost eight hundred feet high at places; it is formed by the upper member of the Hermosa. The total depth of the canyon here is nearly two thousand feet.

August 27, 1968

Orange Cliffs 15' quadrangle, Utah

185

Clearwater Canyon
Camera Station 759

September 27, 1871, morning

This camera station is on the southwest wall of Clearwater Canyon just above its mouth. View is to the northwest, up the canyon. We based the location of the station on parallax between features on the opposite canyon wall, and it is probably accurate to within about five feet. However, because the foreground in Beaman's picture is in shadow, we could recognize few features. Some talus blocks on the opposite canyon wall appear to have moved, but most are in the same places in both pictures. Thor Karlstrom of our expedition posed on the ledge at the top of the falls in the same position as the figure in Beaman's photograph; we can see the reflections of each man in the pool at the base of the falls. The trees on the floor of the canyon above the falls are cottonwoods, three or four of which may be the same in both pictures. The general pattern of water on the floor of the canyon is remarkably similar in the two photographs.

The Cedar Mesa Sandstone Member of the Cutler Formation forms the great alcove facing the camera in the distance, as well as the upper canyon walls. Below it are the Halgaito Member of the Cutler and the Rico Formation. Below them is the upper member of the Hermosa Formation.

August 27, 1968

Mouth of Dark Canyon 15' quadrangle, Utah

September 29, 1871, early afternoon

Cataract Canyon
Camera Station 764

Beaman's camera station was at Mille Crag Bend on the south bank of the Colorado about three quarters of a mile upstream (north) from the mouth of Sheep Canyon. Because his foreground area was covered by about ten feet of Lake Powell, we took our picture from a boat about twenty feet from shore, basing its location on parallax between features on the opposite canyon wall.

The Cedar Mesa Sandstone Member of the Cutler Formation forms the Mille ("One Thousand") Crags on the skyline and the upper wall of the canyon. The Rico Formation forms ledges low on the canyon wall. (The Halgaito Member of the Cutler is not recognizable here.)

188

August 28, 1968

Browns Rim 15' quadrangle, Utah

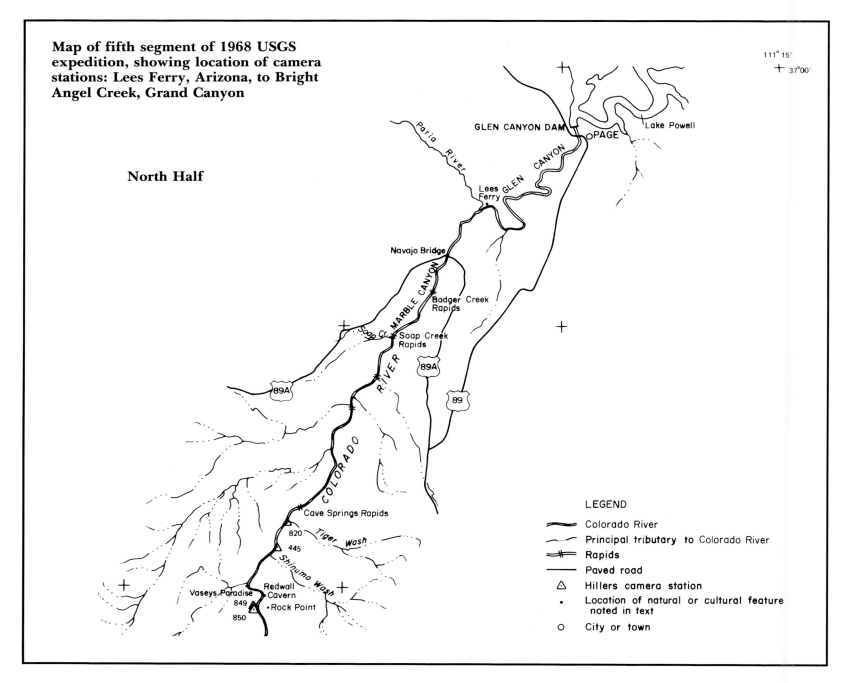

Map of fifth segment of 1968 USGS expedition, showing location of camera stations: Lees Ferry, Arizona, to Bright Angel Creek, Grand Canyon

North Half

111° 15'
37°00'

Paria River

GLEN CANYON DAM
PAGE
Lake Powell

Lees Ferry
GLEN CANYON

Navajo Bridge

MARBLE CANYON

Badger Creek Rapids

Soap Cr.
Soap Creek Rapids

89A

89A

89

COLORADO RIVER

Cave Springs Rapids

820
Tiger Wash

445

Shinumo Wash

Vaseys Paradise
Redwall Cavern

849
Rock Point

850

LEGEND

〰〰 Colorado River
⋯⋯ Principal tributary to Colorado River
✚✚ Rapids
——— Paved road
△ Hillers camera station
• Location of natural or cultural feature noted in text
○ City or town

190

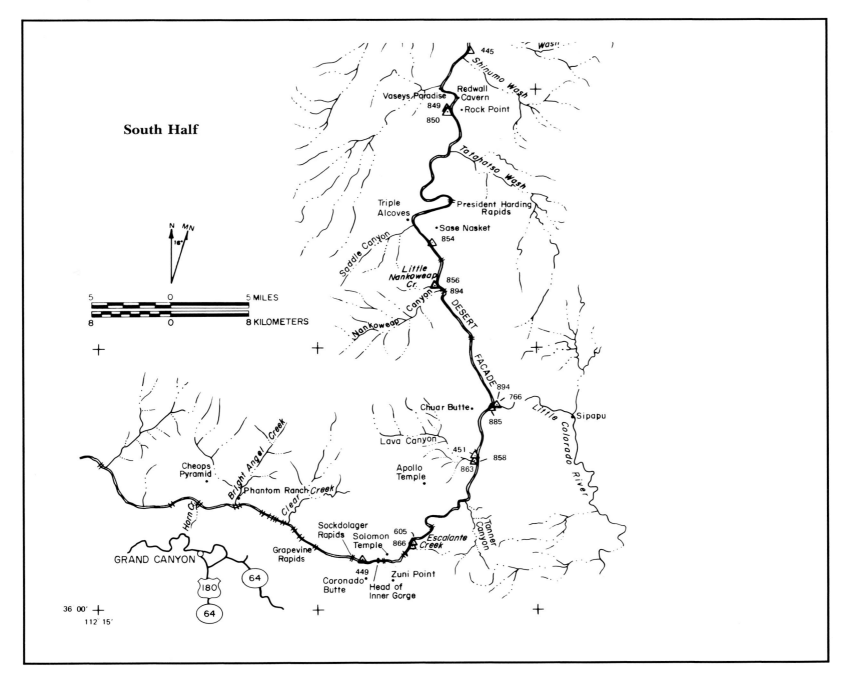

South Half

N MN
16°

5   0   5 MILES
8   0   8 KILOMETERS

445

Wash

Shinumo Wash

Vaseys Paradise
Redwall
Cavern
849   • Rock Point
850

Tatahatso Wash

Triple
Alcoves
President Harding
Rapids

• Sase Nasket
854

Saddle Canyon

Little
Nankoweap
Cr.
856
894

Nankoweap Canyon

DESERT

FACADE

894
766

Chuar Butte •
885

Sipapu

Lava Canyon

Little Colorado River

451
858
863

Apollo
Temple
•

Cheops
Pyramid

Bright Angel Creek

Phantom Ranch
Clear Creek

Horn Cr.

Sockdolager
Rapids
Solomon
Temple   605
866

Escalante
Creek

Tanner Canyon

Grapevine
Rapids

GRAND CANYON

180

64

64

Coronado
Butte •   449
Head of
Inner Gorge

Zuni Point
•

36 00'
112 15'

Table V: Fifth Segment, 1968 Expedition

Geologic formations mentioned in captions of photographs taken in Marble and Grand canyons between Lees Ferry, Arizona, and the mouth of Bright Angel Creek.

Geologic Era	Geologic Period	Formation or Rock Type	Remarks (Numbers are those of Hillers Photographs)
CENOZOIC	QUATERNARY	Holocene talus	Talus of diabase near mouth of Lava Canyon (858).
		Pleistocene talus	Just below Vaseys Paradise (849) and downstream (856, 894, 885).
PALEOZOIC	PERMIAN	Kaibab Formation	Forms uppermost cliff of canyon walls in Marble and Grand canyons; in all photographs as far south as Escalante Creek (866).
		Toroweap Formation } Coconino Sandstone }	In all photographs; form upper canyon walls.
		Hermit Shale	Visible in all photographs as far south as Lava Canyon (858), except in 850, 894, 766, and 885.
	PERMIAN AND PENNSYLVANIAN	Supai Group	In all photographs.
	MISSISSIPPIAN	Redwall Limestone	Prominent as red cliffs extending up from river in Marble Canyon; in all photographs as far south as Escalante Creek (866).
	CAMBRIAN	Muav Limestone } Bright Angel Shale }	Above Little Nankoweap Creek (854) and downriver.
		Tapeats Sandstone	Noted at mouth of Little Colorado (894, 451) and downriver.

192

Geologic Era	Geologic Period	Formation or Rock Type	Remarks (Numbers are those of Hillers Photographs)
MIDDLE PROTEROZOIC		Nankoweap Formation	Near mouth of Lava Canyon (451, 858).
		Cardenas Basalt (Rama Formation of Maxson, 1961)	Near mouth of Lava Canyon (858), where faulted against Nankoweap.
		Dox Sandstone	Near mouth of Lava Canyon (858) and near Escalante Creek (605, 866).
		Shinumo Quartzite	Forms most of Inner Gorge walls near Escalante Creek (866).
		Bass Limestone	Intruded by diabase sills at Sockdolager Rapids (449).
		Hotauta Conglomerate	Basal unit of Unkar that caps Inner Gorge cliff at Sockdolager Rapids (449).
EARLY PROTEROZOIC		Vishnu Schist	Forms Inner Gorge walls at Sockdolager Rapids (449).

(The Formation or Rock Type column units Cardenas Basalt through Hotauta Conglomerate are bracketed together as "Unkar Group.")

Henry Toll takes his inflatable raft through turbulent Soap Creek rapids of the Colorado River in Marble Canyon. Present-day river runners prefer to face downstream, using the oars to maneuver to the right or left.

Chapter Five

Lees Ferry, Arizona, to the Mouth of Bright Angel Creek

Most Grand Canyon river expeditions, both private and commercial, begin at Lees Ferry, about fifteen miles down the Colorado River from Glen Canyon Dam. We left Lees Ferry on September 1, 1968, with five inflatable rafts and ten persons, including five geologists and two photographers. All the rafts were fitted with wooden rowing frames, each equipped with a pair of oarlocks. A spare set of oars was lashed to each boat. The water issuing from the depths of impounded Lake Powell was cold and clear, a haven for large rainbow trout, but it became increasingly turbid as we floated downstream. The canyon walls, low and subdued at Lees Ferry, are steep and towering farther downriver. Just below Navajo Bridge, the red Hermit Shale (Lower Permian) crops out near the river under cliffs of massive Coconino Sandstone fifty to sixty feet high. Cliffs of Toroweap Formation, consisting of interbedded sandstone, mudstone, and cherty limestone, lie above the Coconino, and massive, light-gray, cherty limestone of the Kaibab Formation forms the cliffs at the top. (Details of the geologic formations seen in the canyon walls are given by Simmons and Gaskill, 1969.)

The first major rapids are Badger Creek Rapids, about eight miles downstream from Lees Ferry. Here our party received its baptism in Marble Canyon, when one boat capsized and another bounced a passenger into the frothing white water. We camped on a sandy beach on the east bank about 300 feet below the rapids and spent part of the next day repairing the rowing frame that had splintered during the upset.

Three miles farther downstream we encountered Soap Creek Rapids. Here we suffered another upset when one of our experienced river runners flipped his craft in a deep hole at the upper end of the rapids.

The Powell expedition, having spent the winter of 1871-72 in the settlement of Kanab, Utah, resumed its river voyage down the Colorado on August 17, 1872, with seven men in the *Cañonita* and *Emma Dean*. Jack Hillers was their photographer. (After Beaman left the party during their winter encampment at Kanab, James Fennemore became chief photographer. He was taken ill, however, in the summer of 1872, and his responsibilities passed to his assistant, Hillers, who filled the post with distinction.)

Although we knew that Hillers took several pictures around Lees Ferry, the first camera station in Marble Canyon for which we had photographs was station 820, at the mouth of Tiger Wash. About a mile downstream from Tiger Wash is the site of photograph 445, looking upstream at high canyon walls.

Vaseys Paradise on the west bank is a bright spot of green, where an abundant flow of spring water issues from the Redwall Limestone cliff and tumbles down the slope into the river. We found watercress here, and multitudes of colorful wild flowers.

The Redwall Limestone is the most distinctive formation in Marble Canyon. It forms massive red cliffs, at places more than 400 feet high, that extend for miles along the river's course. Powell, in fact, named Marble Canyon for the outcrops of Redwall polished by the river. Many caves have been formed in it by groundwater flow in Paleozoic and later times. Such a cave is Redwall Cavern, a deep recess at river level that has been formed by recent scouring by the river. The cavern appears in Hillers photograph 849 at mile 33. (River mileages in this segment are measured from Lees Ferry.)

About three miles below Tatahatso Wash (mile 37.6) are the remains of the boat of Bert Loper, an old-time river runner. The boat was pulled up on the right bank just above the high-water mark, and a bronze plaque in Loper's memory has been set in stone near the prow of his home-built boat. Loper lost his life in Tanner Rapids during a run down the canyon in celebration of his 80th birthday. Shoemaker recalled having met him on the Glen Canyon stretch of the river in 1949, when Loper was on his way to rendezvous with friends to begin his Grand Canyon voyage that ended so tragically.

President Harding Rapids (mile 43.7) is recognized by the large solitary boulder splitting the rapids. The remains of Peter Hansborough of the Stanton expedition were found here in 1890 and were buried at the base of a cliff off the east bank. He had drowned in rapids somewhere upstream the previous year. His initials were carved on a rock ledge overhanging the grave. Just above the mouth of Little Nankoweap Creek is the majestic array of the geologic section from Bright Angel Shale at the river to the Kaibab Formation on the rim. Hillers photograph 856 shows this view, as well as details of the *Emma Dean*, which bears Powell's armchair securely lashed to its middle deck, and the *Cañonita*.

We found an excellent campsite on a wide expanse of sandy beach just below the mouth of Little Nankoweap Creek. A hike up Nankoweap Canyon gave us a good opportunity to examine some of the Proterozoic rocks of the Grand Canyon area. We could see prehistoric Indian ruins high in the west canyon wall immediately below the mouth of Little Nankoweap.

The confluence of the Little Colorado and Colorado rivers at mile 61.4 separates Marble Canyon from Grand Canyon. Jack Hillers climbed the intervening ridge to photograph a magnificent scene upriver (photograph 894). We were there to see the evening sun cast a golden light over the red pinnacles and buttresses of the rugged canyon wall upstream. Hillers photograph 766, taken from the same ridge, shows Chuar Butte directly across the river. Camera station 885 was located on an island in the confluence of the two rivers. The shape of the island and the arrangement of its boulders, subject to the whims of both rivers, have changed drastically since 1872.

A four-mile hike up the calcium-charged waters of the Little Colorado brought us to the Sipapu, a travertine mound some thirty feet in diameter and fifteen to twenty feet high, with an opening on the top. A bubbling spring sustains the water level about five feet below the mouth of the opening. Legends of the Hopi tell us that their ancestors emerged from the Earth at this place. Our hike was arduous and wet—we had to wade repeatedly through the pale-blue, milky waters of the river—but our effort was rewarded by the impressive canyon scenes. Nearly a mile and a half below the mouth of the Little Colorado are abundant white crusts of salt on ledges of Tapeats Sandstone overhanging the east bank of the Colorado. Half a mile farther is the unconformable contact between Paleozoic rocks and the underlying Proterozoic Nankoweap Formation; the unconformity represents a major time division in the geologic record. The Nankoweap is shown as low bluffs near river level in Hillers photograph 451, looking upriver.

Half a mile above the mouth of Escalante Creek are two camera sites on the east bank that were occupied by Jack Hillers. One view (605) is upstream toward Apollo Temple. The other view (866), about fifty feet away, is downstream and includes the Shinumo Quartzite overlain by the Dox Sandstone; both of these formations are of Middle Proterozoic age. Other rocks of the canyon walls, extending to Zuni Point, are Bright Angel Shale through Kaibab Formation.

At mile 77.0, the river enters the Upper Granite Gorge, and we first encountered the oldest rock exposed in the Grand Canyon—the Proterozoic Vishnu Schist. The Vishnu, more than 1,700,000,000 years old, makes up most of the Granite Gorge. The Inner Gorge is the more common and accurate name for this part of the canyon because the Vishnu is not granite but is composed of metamorphic rocks—schist and gneiss. These rocks are rich in quartz, feldspar, biotite, and, locally, garnet. They have conspicuous fine layers (foliations) and are commonly faulted and deformed into steeply dipping folds. These dark rocks are cut by abundant dikes of pink pegmatite, which at places roughly parallel the foliation of the enclosing rock. Locally, the dikes have been folded along with the schist and gneiss, resulting in striking patterns of contrasting colors on the canyon walls. At many places near river level, the Vishnu has been highly polished by the silt-laden waters of spring runoffs in the days before Glen Canyon Dam. Because all of these rocks are extremely hard and resistant to erosion, the Inner Gorge is narrow and its precipitous walls plunge directly into the river.

Close to the river on the north bank at mile 77.3 is an abandoned asbestos mine, where serpentine and asbestos occur in the Bass Limestone adjacent to a diabase sill. The dumps from the old workings extend almost to the river's edge. Here the Bass forms the base of the Upper Proterozoic sequence of beds.

We had another excellent view of the steep and confining Inner Gorge from the head of Sockdolager Rapids (Hillers photograph 449). A major unconformity between the top of the Vishnu Schist and the overlying rocks of Late Proterozoic age can be seen on the north wall of the canyon, at the first sharp break in slope on the skyline at the top of the Inner Gorge. ("Sockdolager" is not the name of an early river runner, as one might think, but rather a nineteenth-century slang word for "knockout blow"—an apt choice of name, we decided.)

Grapevine Rapids (mile 81.5) is the largest and most turbulent of the rapids that we encountered during the last few miles of river before reaching the mouth of Bright Angel Creek and the end of segment 5.

August 20, 1872, about noon

Marble Canyon
Camera Station 820

The camera station is at the mouth of Tiger Wash, on the east bank of the Colorado River. View is to the north-northeast, upstream. The step in the skyline near the center of the pictures is a reentrant above Cave Springs Rapids. Precise location of the camera station was based on foreground rocks. A piece of purple cherty limestone breccia with flute marks lies immediately in front of the camera, and a piece of gray limestone, also fluted, lies just beyond it. These and most of the other foreground rocks are still in place, but some large ones along the left side of Hillers's picture have been moved. The boat in his photograph is the *Emma Dean*.

The tamarisks and smaller bushes shown in our picture can grow here today because the river level is controlled by Glen Canyon Dam. Note that it is four to five feet lower in our picture than in Hillers's.

The cliff on the right is formed by the upper part of the Redwall Limestone. Redwall also forms the low bluff on the opposite side of the river and the first series of ledges above the bluff. Reddish-brown sandstone, siltstone, and shale of the Supai Group form the ledges and cliffs extending nearly to the uppermost ledge of the slope above. The next overlying unit is composed mainly of reddish-brown siltstone and mudstone of the Hermit Shale, but the exact location of the contact is difficult to pick.

The cliff extending to the skyline in the background is composed of the upper part of the Hermit, which forms a dark band at the base of the cliff; Coconino Sandstone forms the lower, smooth, light-colored part; Toroweap Formation forms the dark band in the middle; and Kaibab Formation forms the light upper part. No conspicuous changes were observed in the bedrock.

September 3, 1968

Emmett Wash 15' quadrangle, Arizona

199

Marble Canyon
Camera Station 445

August 20, 1872, afternoon

These views were taken from the southeast bank a mile above the mouth of Shinumo Wash, looking upstream. We based the precise location of the camera station on foreground rocks, few of which have been moved since 1872. The water stage in Hillers's photograph is four to five feet higher than in ours.

Redwall Limestone forms the low cliff on the right and the series of ledges and cliffs in the distance on the left as high as the talus slope lit by the sun in 1872. (In Hillers's picture the slopes are in sunlight and the cliffs in shadow, and they are easier to distinguish than in ours; our morning picture shows

September 4, 1968

more detail, however.) The Supai Group and overlying Hermit Shale extend from the base of the talus slope to the foot of a light-colored cliff of Coconino Sandstone and lower Toroweap Formation above. Although not visible from the camera station, cross-bedding in the Coconino is here on a gigantic scale and extends through the entire thickness of the formation. The upper Toroweap forms the next higher slope, and the Kaibab Formation forms the cliff at the top. A relatively fresh appearing rock fall can be seen in Hillers's photograph. No obvious changes were seen in the canyon walls.

Emmett Wash 15' Quadrangle, Arizona

August 21, 1872, about noon

Marble Canyon
Camera Station 849

These two pictures and the pair that follows were taken about 500 feet apart from the west bank of the river opposite Rock Point (out of sight to upper right). This view is north-northeast, looking upstream toward Redwall Cavern (visible at the far bend of the river). Hillers's noon picture enabled him to capture details on the far canyon wall that we were unable to photograph, because we were there in late afternoon when the wall was bathed in brilliant sunshine. We based the camera station location on parallax between features on the canyon walls and on large rocks along the left side of the picture, and it is probably accurate to about a foot, although the water stage was at least six feet higher in the Hillers photograph. A pyramid-shaped boulder in the foreground is in the same place but has been rotated since 1872. A few tamarisks are now scattered along both banks of the river. Less windblown sand is present at the foot of talus slopes on the left side of our picture, but more sand has accumulated in the small cave on the right.

Redwall Limestone forms the steep cliffs rising from both sides of the river. The Supai Group forms the steep ledgy cliff that extends almost halfway to the rim. Above it is a slope of uppermost Supai and Hermit Shale; the slope is largely talus covered. Coconino

202

Sandstone forms the next higher cliff, surmounted by buttresses of lower Toroweap Formation. (The buttresses cast long shadows in our picture.) The skyline cliff is Kaibab Formation with upper Toroweap at the base. Most of the talus in sight is probably late Pleistocene; all of it is partly dissected. Redwall Cavern and some of the other caves in view are formed in the upper part of the Redwall Limestone.

September 4, 1968

Nankoweap 15' quadrangle, Arizona

Marble Canyon
Camera Station 850

This camera station is on the west bank of the river about 500 feet downstream from the previous one, looking south-southeast (downstream). Hillers's view includes the *Cañonita* (foreground), the *Emma Dean*, and the darkroom tent (lower right) in which he processed his photographic plates. Precise identification of the nearest foreground rocks in Hillers's picture is uncertain because in 1872 they were partly covered by sand or the river and partly obscured by the darkroom tent. However, we located the camera station on the basis of rocks in the middle distance and on parallax between the near canyon wall and more distant points. Its location is probably accurate to within about two feet, the principal error being along the line of sight.

The lower canyon walls are Redwall Limestone. The large rock protruding from the water behind the *Cañonita* is limestone breccia from the upper part of the Redwall. Supai Group caps the more distant point framed by the canyon walls. The skyline cliff includes the Coconino, Toroweap, and Kaibab, as described for the previous camera station. No changes were observed in the canyon walls.

August 21, 1872, about noon

September 5, 1968

Nankoweap 15' quadrangle, Arizona

August 22, 1872, late morning

Marble Canyon
Camera Station 854

Were it not for the aging of the catclaw acacia tree near the river bank, and the clouds that 1872 photography could not capture (if indeed the clouds were present), it would be difficult to say which of these pictures is Hillers's and which is Stephens's. Both were taken about 11:30 A.M. The camera station is on the northeast river bank below the Sase Nasket (a Navajo term for a point on the canyon rim), one-and-a-half miles below the mouth of Saddle Canyon. The view is upriver to the northwest, toward the Triple Alcoves, the uppermost of which appears in the center of the pictures. Precise location of the camera station was based on foreground rocks.

The catclaw acacia and the tree on the right are still alive, and the same species of cactus—Utah agave, barrel cactus, and prickly pear—are growing on the foreground slope. Most of the sandbank vegetation in the distance is mesquite and, in 1968, tamarisk and a few hackberry trees.

Rocks of the canyon walls range from Bright Angel Shale of Cambrian age to Kaibab Formation of Permian age. Bright Angel forms the low bluffs just above the river and the slope extending to the first high cliff on the right. The lower fourth of this cliff is Muav Limestone, also Cambrian, and the upper three-fourths is the Redwall Limestone

206

of Mississippian age. Their contact cannot be picked out in the photographs, but it is a disconformity representing a time interval of at least 140 million years. On the distant canyon wall facing us, framed between and above Muav-Redwall cliffs, are ledges and slopes of the Supai Group and a slope of Hermit Shale, which is partly covered by talus. This slope is surmounted by a bold cliff of Coconino Sandstone capped by basal Toroweap Formation. Above this cliff is another major slope, also Toroweap. The skyline cliff is Kaibab Formation.

The large rocks in the foreground are talus blocks that have tumbled from the Redwall cliff. The largest block, on the left, is limestone banded by thick lenses of chert. The canyon walls and the sand beach and boulder fan on the opposite side of the river are about the same today as in 1872.

September 6, 1968

Nankoweap 15' quadrangle, Arizona

August 22, 1872, early afternoon

Marble Canyon
Camera Station 856

This camera station is on the west bank of the river just above the mouth of Little Nankoweap Creek. The view is slightly east of north, upstream. Location of the camera station was based on parallax between features on the canyon walls; its accuracy is within about five feet perpendicular to the line of sight and within about ten feet along the line of sight.

Bright Angel Shale forms the low bluffs rising above the river. On the right it is largely covered by dissected coarse Pleistocene talus. Muav Limestone at the base of the cliff above the Bright Angel is prominent as buttresses standing out from the cliff (in sunlight in Stephens's picture). The main part of the cliff is Redwall Limestone. The Supai Group forms ledges and slopes above the Redwall cliff visible on the right side of the pictures. Hermit Shale and the uppermost part of the Supai form a higher slope partly covered by talus. The skyline cliff on the right includes the Coconino Sandstone and the Toroweap and Kaibab Formations.

At least one fresh spall appears to have formed since 1872 in the alcoves in the Redwall near the center of the pictures. The sand beach on the right is now fringed with tamarisk, but mesquite is growing at its top at the same level as in 1872.

September 6, 1968

Nankoweap 15' quadrangle, Arizona

August 23, 1872, afternoon

pattern of larger rocks on the fan is
almost the same today as in 1872, although there is now less windblown
sand. The bushes along the right river
bank in both pictures are mesquite, and
tamarisk is now scattered along both
sides of the river.

September 8, 1968

Vishnu Temple 15' quadrangle, Arizona

211

August 22, 1872, afternoon

Marble Canyon
Camera Station 766

Still on the east bank, we have now moved slightly up the same ridge on which the previous camera station was located and have turned to face due west. We see the island, now partly submerged by high water, at the confluence of the Colorado and Little Colorado rivers and also the fully majesty of Chuar Butte.

When we reached this point in 1968, we did not yet have the Hillers photograph, only a line drawing based on the photograph; the drawing had appeared in the Powell report of 1875. The close correspondence of the two pictures shown here can be attributed to the accuracy of that drawing.

The rocks in view are the same as those shown in the previous pair of photographs: Tapeats Sandstone forms the low ledge above the water. The slope and intermittent ledges rising above this ledge are Bright Angel Shale, topped by a distinct ledge of Muav Limestone. Redwall Limestone forms the great cliff above this ledge (in about the center of the entire section as seen from the river). The Redwall cliff is capped by a series of Supai Group ledges and slopes. The high cliff above is Coconino Sandstone and Toroweap and Kaibab Formations. Toroweap and lowermost Kaibab form the jagged fingers on the highest part of the distant skyline.

September 8, 1968

Vishnu Temple 15' quadrangle, Arizona

Marble Canyon
Camera Station 885

August 22, 1872, afternoon

The camera station is at the confluence of the Colorado and Little Colorado rivers, on the island shown in the previous pictures. Our view is to the north, up the Colorado River toward the Desert Facade. Our location of the camera station was based on parallax between features on both banks of the river and is probably accurate to within a foot, although we can see that the island has changed considerably. Its northeast side now extends fifty feet farther upstream in the direction of the line of sight.

Rocks along the canyon wall range from upper Tapeats Sandstone to Kaibab Formation. Tapeats forms the low ledges at water level along the extreme right side of Hillers's picture. Just above and to the left of these ledges is a large mass of cemented Pleistocene talus. Details of the canyon wall are given in the preceding captions (for photographs 894 and 766).

September 8, 1968

Vishnu Temple 15' quadrangle, Arizona

Grand Canyon
Camera Station 451

August 26, 1872, late morning

Our view here is from the west bank of the river half a mile above Lava Canyon. We are looking slightly east of north, upriver. We located the camera station precisely on the basis of foreground rocks, most of which are slabs of crumbly sandstone that have fallen from a cliff of Nankoweap Formation. The larger rocks are in the same positions in both pictures. Immediately in front of our camera, frag-ments have fallen onto the flank of a small gully, but other material has been washed out from the gully itself, requiring that we build the site up in order to occupy the camera station.

The Nankoweap Formation of Proterozoic age forms the low cliffs and bluffs along the river on the right and in the distance. The Nankoweap is trun-cated by an angular unconformity at the base of the

216

September 9, 1968

Cambrian Tapeats Sandstone. The Tapeats forms cliffs low on the canyon walls on both sides of the river. Upstream, framed between the Tapeats cliffs, is a slope of Bright Angel Shale partly covered by talus and surmounted by a ledge of Mauv Limestone. Above them rises the familiar cliff of Redwall Limestone. Supai Group forms a series of very steep slopes and ledges above the Redwall near the center of the pictures. The Hermit Shale forms a lighter colored, talus-covered slope overlain by a rather broken, ragged, light-colored cliff of Coconino Sandstone that grades up into a cliff of Toroweap and Kaibab Formations. No changes were noticed in the canyon walls.

Vishnu Temple 15' quadrangle, Arizona

August 26, 1872, noon

Grand Canyon
Camera Station 858

This camera station and the next are about a thousand feet upstream from the mouth of Lava Canyon and 1,700 feet downstream from the previous station and are also on the west bank. The view is south-southeast, across and down the river. Our precise location of the camera station was based on the foreground outcrop of dipping sandstone and shale. These beds are in the lower part of the Nankoweap Formation and were tilted along a fault that placed them in contact with Proterozoic diabase (the Cardenas Basalt), seen across the river on the right. There the fault trace can be seen (in shadow in our picture) just behind a blade of rock above the boulder fan. Powell's party found copper and silver minerals here, which a prospector, Tanner, later mined.

Across the river in the distance, exposed rocks range from the Middle Proterozoic Dox Sandstone to the Permian Kaibab Formation on the canyon rim. Thin, red, shaly sandstone beds of the Dox can be seen close to the river just to the left of the foreground wall (on which we can see ripple marks in Hillers's picture). Closer to us, these thin Dox beds are partly covered by talus. The peak in the center of the pictures is a great mass of diabase resting on the Dox. To the left of this peak, above another talus fan, the Nankoweap Formation forms a low ridge. (The Nankoweap is younger than the diabase but has been downdropped at the fault noted above.) The diabase peak is capped by Tapeats Sandstone. Behind and above the Nankoweap ridge is a cliff of Redwall Limestone and, in ascending succession, the Supai, Hermit, Coconino, Toroweap, and Kaibab. A great alcove near the left edge of the pictures was formed in Coconino, and its roof is Toroweap.

The far river bank is much the same today except that willow and tamarisk now extend down to the water. Mesquite and catclaw acacia are growing farther back in both pictures. Hillers's darkroom tent appears in his foreground. As we can see, he had difficulty finding a level spot for it.

September 9, 1968

Vishnu Temple 15' Quadrangle, Arizona

August 26, 1872, noon

Grand Canyon
Camera Station 863

This camera station is about the same as the previous one (858), and Hillers probably took his pictures at the same time. View is upstream. We have no matching 1968 picture because, when we rephotographed the previous picture, we were not aware that Hillers had also taken this upriver scene. By the time we realized our oversight, we were two miles downstream!

The foreground and low cliffs on both sides of the river are Nankoweap Formation. Tapeats Sandstone forms the next higher cliff. The younger rocks are as described in the previous caption.

Vishnu Temple 15' quadrangle, Arizona

220

August 28, 1872, morning

Grand Canyon
Camera Station 605

We are here on the southeast bank, looking upstream slightly east of north, toward Apollo Temple. We based the location of our camera station on parallax between features on near and distant canyon walls and on the temple. However, we could recognize no features on the sand bank in the foreground, and our location may be in error by as much as twenty feet along the line of sight. The large rocks on the bank in the middle distance are not recognizable in Hillers's picture. They may have fallen from the cliff since 1872, or they may have been hidden in the dense growth of mesquite.

The rocks of the lower canyon walls are all part of the Dox Sandstone. They include a lowermost dark-brown sandstone and a higher deep-red sandstone, which forms the bench framed between the lower canyon walls in the middle distance. The rocks in view on Apollo Temple range from Tapeats Sandstone to lower Supai Group. Tapeats forms a low cliff in front of the temple. The ledgy slope on the right flank of the temple is Bright Angel Shale. Muav and Redwall Limestones form the high cliff and also the spire on the right, which has been downdropped by faulting. Lower Supai caps the temple.

222

September 10, 1968

Vishnu Temple 15' quadrangle, Arizona

August 28, 1872, morning

Grand Canyon
Camera Station 866

This camera station is about fifty feet downstream from the previous one, also on the southeast bank, and about half a mile upstream from the mouth of Escalante Creek. The view is to the southwest, downriver. Zuni Point is on the extreme left, Coronado Butte in the center, and the foot of Solomon Temple on the right. The location of the camera station was based on foreground rocks, but we were unable to occupy its exact position because the sand bank on which Hillers stood has been partly washed away. By raising the camera tripod to its maximum height, however, Stephens was able to get within a few inches of the correct position.

Most of the largest foreground rocks are still in place, but some changes have occurred since 1872. A large sandstone boulder has rolled or been washed down against the two biggest rocks in the center of Hillers's picture, forcing aside the one nearest the center and rotating it. In Hillers's picture, this rock was already cracked along the top, and a big slab has fallen from it toward the river. The distinctive sharp tip of the rock nearest Hillers's camera has also been broken off.

White, red, purple, and brown sandstone of the Shinumo Quartzite of Middle Proterozoic age forms the head of the Inner Gorge in this view; the conspicuously banded cliff is the upper-

224

most part of the Shinumo. These beds are dipping about 15° toward the camera. Lowermost Dox Sandstone forms the benches above. At the left, Zuni Point consists of formations ranging from upper Bright Angel Shale through Kaibab Formation. The prominent cliff low on Zuni Point is Redwall Limestone. Behind it, in the far distance, the lighter colored slope is Coconino Sandstone and Kaibab of Moran Point. The beds in view on Coronado Butte in the central background include Coconino, Toroweap, and possibly lowermost beds of Kaibab.

September 10, 1968

Vishnu Temple 15' quadrangle, Arizona

225

Grand Canyon
Camera Station 449

August 29, 1872, about noon

We are now on the north river bank at the head of Sockdolager Rapids. The view is to the west, downstream. Framed between the walls of the Inner Gorge in the middle distance is the abutment of Newberry Butte.

Precise location of the camera station was based on foreground rocks, almost all of which are present and accounted for. All have polished faces due to abrasion by sediments suspended in the river at high-water stages. The huge boulder of which we see a part along the right side of the pictures is pegmatite. Just below it and slightly to its left is a lighter banded rock of pegmatite and schist. Another such block occurs along the bottom left edge of Hillers's picture, and still another is at the extreme left. Three prominent rocks close together in the middle of the photographs near the camera, including the one that hides part of the river, are the Hotauta Conglomerate.

September 11, 1968

The walls of this part of the Inner Gorge are formed in the Vishnu Schist, here cut by numerous dikes and ribbons of pegmatite. Some of these steeply dipping rocks can be seen about end on at the water's edge across the canyon at the left edge of the pictures. On the right canyon wall, the great unconformity between the Vishnu Schist and the overlying Unkar Group occurs at the base of a ledge of Hotauta that caps the Inner Gorge cliff. Above the ledge is a slope formed on the Bass Limestone and on a diabase sill intruded near the base of the Bass. On the left canyon wall, the Hotauta is just barely visible along the highest part of the skyline. The ragged part of the skyline on the left is a more distant cliff formed by the Tapeats Sandstone. In the haze in the distance is a promontory with a small peak, both formed of the Supai Group. Just to the left of this peak, a cliff of Coconino is capped by Toroweap. No changes were detected in the canyon walls.

Vishnu Temple 15′ quadrangle, Arizona

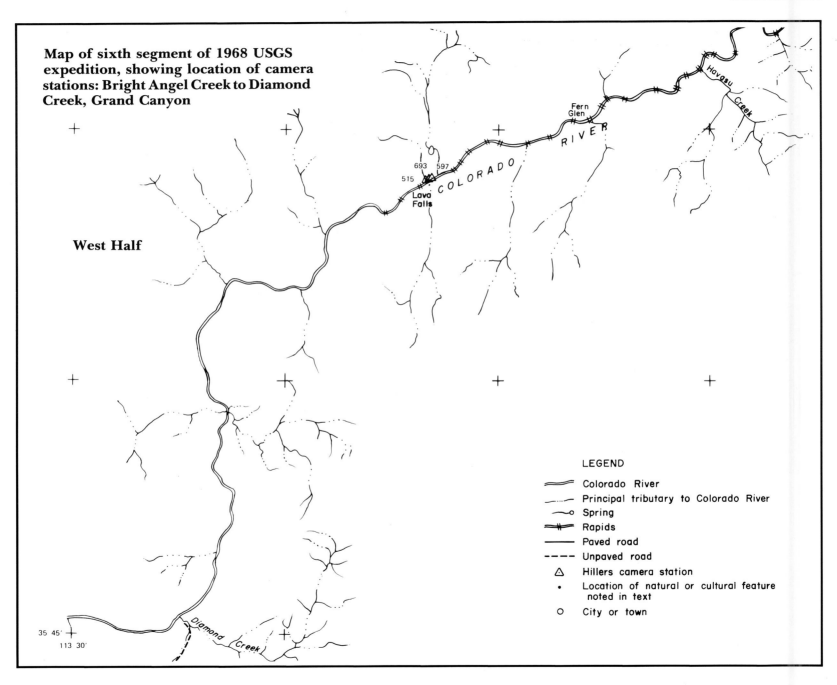

Map of sixth segment of 1968 USGS expedition, showing location of camera stations: Bright Angel Creek to Diamond Creek, Grand Canyon

West Half

LEGEND

Colorado River
Principal tributary to Colorado River
Spring
Rapids
Paved road
Unpaved road
△ Hillers camera station
• Location of natural or cultural feature noted in text
○ City or town

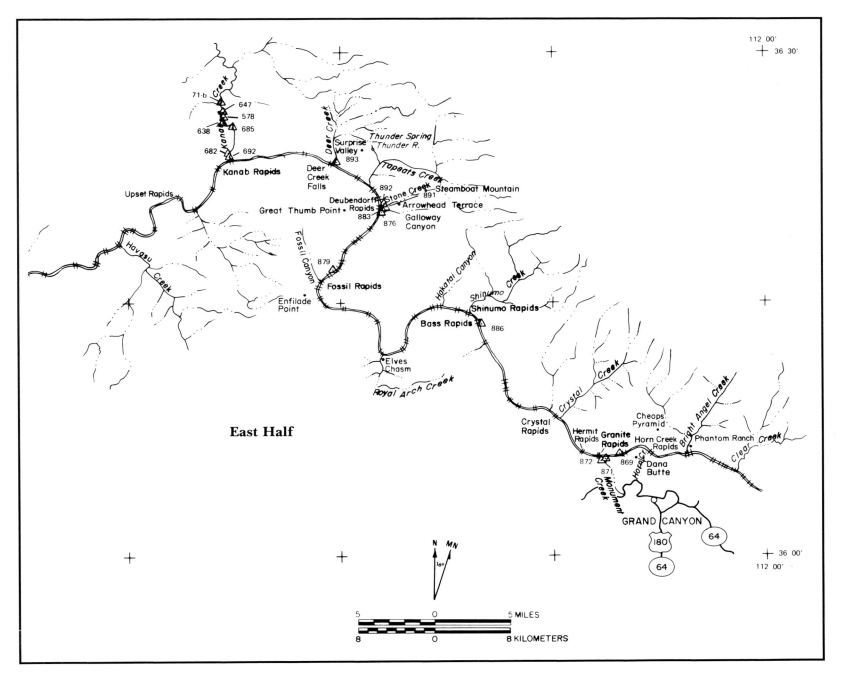

East Half

229

Table VI: Sixth Segment, 1968 Expedition

Geologic formations mentioned in captions of photographs taken in the Grand Canyon, Arizona, between the mouths of Bright Angel Creek and Diamond Creek.

Geologic Era	Geologic Period	Formation or Rock Type	Remarks (Numbers are those of Hillers Photographs)
CENOZOIC	QUATERNARY	Pleistocene talus and colluvium	Near Deubendorff Rapids (876); in Kanab Creek canyon (692, 685); at Lava Falls (515).
		Pleistocene fans and basalt flows	At Lava Falls (597, 693).
PALEOZOIC	PERMIAN	Kaibab Formation	Forms uppermost cliff of canyon walls; in pictures between Bass Rapids (886) and Tapeats Creek (892).
		Toroweap Formation } Coconino Sandstone	Noted where upper canyon walls are visible in pictures; in captions as for Kaibab (above).
		Hermit Shale	Near Bass Rapids (886) and downstream (879, 892).
	PERMIAN AND PENNSYLVANIAN	Supai Group	In most photographs downriver from Bass Rapids (886) and along Kanab Creek.
	MISSISSIPPIAN	Redwall Limestone	Seen in almost all photographs.
	DEVONIAN	Temple Butte Limestone	Fills channels on eroded surface of Muav Limestone. Occurs near Bass Rapids (886) and downriver, and along Kanab Creek.
	CAMBRIAN	Muav Limestone Bright Angel Shale } Tapeats Sandstone	Noted in almost all captions.

230

Geologic Era	Geologic Period	Formation or Rock Type	Remarks (Numbers are those of Hillers Photographs)
MIDDLE PROTEROZOIC		Unkar Group { Shinumo Quartzite	Near Granite Rapids (872) and Bass Rapids (886).
		Hakatai Shale	Near Bass Rapids (886) and Deubendorff Rapids (876, 883, 892).
		Bass Limestone	Near Deubendorff Rapids (876, 883, 891, 892).
EARLY PROTEROZOIC		Vishnu Schist	Forms Inner Gorge walls near Granite Rapids (869, 871, 872) and Bass Rapids (886).

Group portrait taken at our last camp in Grand Canyon. Left to right: Hal Stephens, Maurice Brock, George Ogura, David Gaskill, George Simmons, Henry Toll, and Gene Shoemaker. Gene wears the decoration awarded him for being dunked in Lava Falls.

Chapter Six

The Mouth of Bright Angel Creek to the Mouth of Diamond Creek

P hantom Ranch, half a mile up Bright Angel Creek from the river, is the place to change crew members and to take on supplies for an extended river trip through the Grand Canyon. The ranch offers meals, lodging, and telephone and radio communication with the outside world. Food supplies can be brought down from the South Rim on mules by way of the Kaibab Trail if prior arrangements have been made.

After taking advantage of the amenities of the ranch, we commenced segment 6 of the river expedition with eleven persons and the same five boats. We started at Bright Angel Creek (mile 87.6 from Lees Ferry). Here a major structure, the Bright Angel fault, crosses the Grand Canyon from rim to rim; Bright Angel Canyon follows it northeastward from the river.

Rapids below Glen Canyon Dam may change their character from time to time, depending on the discharge of water from the dam. Each major rapid should be examined carefully from the nearby bank before proceeding through. At Horn Creek Rapids, less than three miles below Bright Angel Creek, a boat and crew normally glide down the smooth tongue in the swiftest part of the rapid, ride over high standing waves where the tongue narrows to a point, and then continue over the choppy lower waves below. In 1968, however, we had to skirt the left side of the tongue to avoid holes directly ahead of it and then pass close to one large hole and along the edge of a prominent plume of water near the center of the channel, staying well to the left. This course avoided most of the churning waters of the deep holes below the tongue. On a different day, another strategy might be required. A flash flood from a side canyon or a rock fall might produce a marked change in any set of rapids.

In the Inner Gorge, foaming, turbulent rapids are separated by long stretches of calm, smoothly flowing water. The quiet of serene reflecting pools and smooth flows is frequently counteracted by the dull, booming roar of new rapids just around the next bend. The steep walls of the gorge are dark-brown to black schist and gneiss of the Vishnu Schist cut by many light-tan to pink pegmatite dikes, some very thick: one giant is more than 200 feet across. Thousands of interesting forms on the canyon walls and in the side canyons are sculptured by previous high-water episodes. At places, the walls are undulatory and highly polished and resemble draperies.

A fine campsite on a flat sandy bench, protected by large boulders against the canyon wall and shielded by tamarisk, awaits the river traveler just above Granite Rapids at mile 93.4. A fan of coarse, rocky debris derived from Monument Creek, coming in from the south, is responsible for these rapids. At their foot,

Hillers set up his camera among the boulders and made photograph 871, looking upriver, in the early afternoon of September 1, 1872.

Hermit Rapids, less than a mile-and-a-half downstream, offers one of the most exhilarating rides on the river. Standing waves are so high that boats disappear from sight as they pass from crest to trough when viewed from the boulder bar near the foot of the rapids.

Crystal Rapids, at mile 98.1, has been dangerous since 1966, when a flash flood in Crystal Creek deposited a new debris fan that crowded the river against the southwest wall of the canyon. The steep descent of the river at this constriction results in a deep hole and a high standing wave at the foot of the drop. We suffered a spectacular one-boat upset in Crystal, resulting in the loss of a fine camera and the splintering of one rowing frame and oar, but no one was injured.

On Shinumo Creek, less than 300 feet from its mouth, is a beautiful waterfall enclosed on three sides by sixty-foot cliffs of polished Vishnu Schist. Many rainbow trout have been caught in the pool beneath the falls and in the creek above.

River-running parties enjoy the stop at Elves Chasm (mile 116.5). This lovely, shady side canyon has a series of waterfalls, grottoes, and quiet pools surrounded by ferns and moss. A special treat for those just leaving Elves Chasm is the view downriver: a tower of green Bright Angel Shale stands majestically before a great curving amphitheater carved in the Redwall Limestone. The vast, sheer walls of the canyon overwhelm the viewer in this spectacular part of the canyon.

The unconformable contact between Tapeats Sandstone of Cambrian age and Early Proterozoic gneiss comes within fifty feet of the river as one glides downstream from Elves Chasm. One mile below Fossil Rapids (mile 124.7), a Hillers camera station (879) was located on a most unusual boulder fan. Boulders of sandstone and limestone have become interlocked in places where vibration due to turbulence in the river has ground them one against another at times of high water. The incessant rubbing of the softer and harder rocks together has caused the softer ones to wear away and the more resistant ones to penetrate them.

More Hillers camera stations were located at Deubendorff Rapids (876) at mile 131.6, at the mouth of Galloway Canyon (883), and at Stone Creek canyon (892 and 891), two-tenths of a mile farther downstream. About two miles below Stone Creek canyon, Tapeats Creek comes in from the north. Here we found one of the best campsites on the river; the permanent stream provides not only drinking water but many trout. A two-mile hike up the scenic Tapeats Creek trail leads to the mouth of Thunder River. The river—indeed it is tributary to the creek!—issues from a cavern high in a cliff of Muav Limestone, and it bounces down the canyon wall in a series of falls and cascades. Trails from the confluence of the two streams lead out of the canyon to points on the North Rim.

Deer Creek Falls, about 150 feet up Deer Creek (mile 136.2), is the subject of Hillers photograph 893. He set up his camera station on a promontory to record the 100-foot fall of the creek as it flows out of a crevice in the overhanging wall. The foreground rocks, unchanged since Hillers's day, are slabs of Tapeats Sandstone fallen from the cliff above.

A few miles downstream is the mouth of Kanab Creek (mile 143.5), where the Powell expedition ended their river trip and left the two boats, the *Cañonita* and *Emma Dean*, behind. They were met here by George Adair, Joe Hamblin, and Nathan Adams, who had come from Kanab, Utah, with horses and fresh rations. The meeting was described in Hillers's diary (Fowler, 1972).

On September 10, 1872, Jack Hillers and Clem Powell began to take a series of pictures along Kanab Creek on their way out to the town of Kanab. (Other members of the party, including Major Powell, had left for Kanab the day before.) The seven pictures included here are among those taken over a distance of about eight miles from the mouth of the creek.

We continued down the river to Lava Falls (mile 179.3) to recover three camera stations occupied in April 1872 by James Fennemore and Hillers, who was then Fennemore's assistant. They had climbed down the lava-covered slopes from the North Rim and had made several exposures of Lava Falls, generally considered to be the most hazardous of all the remaining rapids in the Grand Canyon. They also photographed the canyon from Toroweap on the North Rim. Their journey was the only one made by members of the Powell party to the lava slopes and Lava Falls far below.

We officially ended our river trip at the mouth of Diamond Creek, which is accessible by gravel road from Peach Springs and U.S. Highway 66. Some of the river party left here, but the rest of us continued to Pearce Ferry on the upper end of Lake Mead, because washouts on the Peach Springs–Diamond Creek road are an ever-present possibility; therefore, the problems of transporting the inflatable rafts and supplies back to Flagstaff from Diamond Creek could have been enormous. We had spent three months traveling nearly a thousand miles to retrace Powell's historic river voyage.

September 1, 1872, noon

Grand Canyon
Camera Station 869

The camera station is on the north bank of the river in the Inner Gorge about five-and-a-half miles below the mouth of Bright Angel Creek. The view is east-northeast, upriver. Hillers, taking his picture under optimum lighting conditions, was able to avoid the afternoon shadows present in our picture. Precise location of the camera station was based on the position of foreground rocks, which are great blocks of pegmatite that have tumbled from the cliffs above the camera station.

Very steeply dipping to vertical foliated metamorphic rocks of the Vishnu Schist form these canyon walls. (The Vishnu is described in the latter part of the introduction to Chapter 5.) The banded appearance of the walls on the left is due to abundant pegmatite dikes, most of which have been squeezed into tight folds along with the enclosing rocks. Tapeats Sandstone forms the ledge capping the spur on the right, which is just below Dana Butte. Bright Angel Shale locally forms a slope along the skyline above this ledge and also most of the slope at the foot of Cheops Pyramid, framed by the Inner Gorge walls. The Muav and Redwall Limestones make up the pyramid itself; the Muav forms a distinct dark ledge at the foot of the Redwall cliff. No conspicuous changes were noted in the canyon walls.

236

September 15, 1968

Bright Angel 15' Quadrangle, Arizona

September 1, 1872, afternoon

Grand Canyon
Camera Station 871

We are now on the south bank at the foot of Granite Rapids, looking upriver across the head of the rapids. Powell's camp no. 100 is in the foreground. The *Cañonita* and the *Emma Dean* have been hauled out and overturned for repairs, and part of Hillers's darkroom tent can be seen at the left. We based the precise location of the camera station on foreground boulders, which are part of the fan of Monument Creek that is responsible for the rapids. Only a few rocks have been moved—some at the lower left and one large one at the right edge of Hillers's picture that hides part of the head of the rapids. This rock has been rotated and has slid downslope. Our water level was at least six feet lower than Hillers's, and we could see many more boulders.

The point on the central skyline is at the foot of Dana Butte. This point and other skyline features look much more eroded in 1872 than in 1968! The "erosion" is due to the early retouching of the original negative with opaque material to remove blemishes in the emulsion. The true skyline must have been nearly identical to that shown in the unretouched photograph taken by Stephens in 1968.

The Vishnu Schist that forms the walls of the Inner Gorge here dips very steeply to vertically. Some pegmatite dikes are conspicuous in the canyon walls at the left. One of these dikes, just above the falls, is about fifty feet across. In the center of the pictures is the Tapeats Sandstone ledge that caps the cliffs of the Inner Gorge; above it is a ledgy slope of Bright Angel Shale. The point on the central skyline is formed in the lower and middle parts of the Redwall Limestone. At the base of this cliff is a distinct ledge of Muav Limestone.

September 16, 1968

Bright Angel 15′ quadrangle, Arizona

September 1, 1872, afternoon

Grand Canyon
Camera Station 872

This camera station is about 900 feet southwest of the previous one, still on the south bank just below the mouth of Monument Creek. The view is east-northeast, upcanyon. The mouth of Monument Creek and part of the head of Granite Rapids are visible in the lower left, but the much lower water level in our picture gives the falls a far different aspect. In addition, there has evidently been a flash flood in Monument Creek since 1872, which has changed the stream bed and deposited new boulders on the debris fan. Powell's camp no. 100 was just beyond the lower left of Hillers's picture. Precise camera-station location was based on details of a pegmatite dike that forms a ridge on the west side of Monument Creek canyon; the ridge is just visible along the bottom edge of the pictures. A cracked microcline crystal (a type of feldspar) a few inches across at the bottom center of the Hillers picture was one of the features used to locate the camera station, but it and the other foreground features were so close to Hillers's camera and were so small that we recognized them only with difficulty. Once we found the precise site, we could identify other individual crystals in the pegmatite in Hillers's photograph.

Most of the rocks seen in this pair of photographs are Vishnu Schist. One of the largest pegmatites in the canyon, one more than 200 feet across, forms the point near the center of the pictures and the profile of the left canyon wall. Above the Vishnu in the right middle distance is a ledge of Tapeats Sandstone capping the Inner Gorge. A long slope of Bright Angel Shale rises above the ledge. The southern spur of Cheops Pyramid in the distance includes rocks ranging from Shinumo Quartzite up to the Redwall Limestone. The Muav and Redwall form the pyramid; below it, Bright Angel Shale forms a slope with one prominent ledge. The Shinumo forms a high cliff at the foot of this slope that is only dimly visible in Hillers's picture. The Shinumo here formed a hill rising out of a peneplain that became an island in the Cambrian sea at the time the Bright Angel was laid down.

Our retouching of four conspicuous cracks in Hillers's plate was not entirely successful. They can be seen splaying out to the left from a point high on the right bluff.

September 16, 1968

Bright Angel 15' quadrangle, Arizona

241

September 3, 1872, noon

Grand Canyon
Camera station 886

Hillers's camera station here was on the east bank of the river a third of a mile above Bass Rapids. The view is to the northwest, downriver. The 1872 water level was at least ten feet higher and the edge of the water about fifty feet closer to the station than when we were there. (The river is visible in our picture only at the extreme left.) Precise location of the camera station was based on foreground ledges of diorite intruded into the Vishnu Schist. Details of these ledges are almost the same as in Hillers's picture. Much of the sand in Hillers's immediate foreground, however, has been eroded by wind. Although we were not able to bring the boats into the field of view, G. G. Anderman posed near the place of the figure sitting just to the right of Powell's boats. Four catclaw acacia trees growing in the center and right foreground of the 1872 picture are still alive and seem to be about the same size and shape.

All the rocks in the foreground and those on the low, dark cliff at the left are Early Proterozoic diorite. The diorite is cut by a few thin pegmatite dikes, one of which can be seen as a thin, irregular, light streak near the lower right side of the 1968 picture. Downcanyon, the Vishnu is overlain by the Unkar Group of Middle Proterozoic age. In Stephens's picture, the

242

dark rocks low on the slope near the center are part of a thick diabase sill intruded into the Hakatai Shale; the first tilted beds overlying the dark rocks are Hakatai.

Shinumo Quartzite can be seen as peaks on the left in front of the distant canyon wall and at the skyline on the right. In the distance is a series of cliffs and ledges of Paleozoic rocks that form a backdrop to exposures of the Unkar Group. Bright Angel Shale forms the slope behind the peaks of Shinumo Quartzite. The first thick ledge above this slope is Muav Limestone. Devonian Temple Butte Limestone may form an indistinct ledge above the Cambrian Muav. Redwall Limestone makes up prominent buttresses and cliffs, and the Supai Group forms a conspicuous buttress extending to the skyline on the right. Hermit Shale, Coconino Sandstone, and Toroweap and Kaibab Formations form a slope-cliff-slope-cliff sequence on the flank of the Masonic Temple along the central skyline.

September 18, 1968

Havasupai Point 15' quadrangle, Arizona

243

September 5, 1872, morning

Grand Canyon
Camera station 879

We are now much farther downriver, on the northwest bank about a mile below Fossil Rapids. The view is to the southwest, upriver, toward the mouth of Fossil Canyon and Enfilade Point. The standing figure of 1872 is identified in Hillers's album as Thompson; our figure is George Ogura. The river is about fifteen feet higher in the 1872 photograph than in ours. (In the Hillers album, a picture taken twenty feet north of this one was mislabeled Marble Canyon, and a line drawing based on it in Powell's Smithsonian report was misidentified as "Head of Grand Canyon." The geology and general scene do, in fact, resemble those in part of Marble Canyon just above the mouth of the Little Colorado River.)

Tapeats Sandstone forms the ledgy slope and low cliff just above the river. The slopes rising above them, and the intervening ledge, are Bright Angel Shale. Muav Limestone and possibly Temple Butte Limestone form the next main ledge. Above them is a Redwall Limestone cliff with a great alcove on the left. Supai Group forms a slope and a series of cliffs and ledges above the Redwall. Conspicuous benches are formed on sandstone beds high in the Supai. Behind these benches is the cliff-slope-cliff sequence of Coconino, Toroweap, and Kaibab on Enfilade Point near the center of the pictures.

Location of the camera station was based on foreground boulders, most of which are still in place, although at least two large ones are missing. Just in front of the camera station we discovered the most remarkable example of interlocking boulders that any of us had ever seen: one pink sandstone boulder is closely molded to at least five other boulders. The fitting together of the boulders is apparently due to vibration or agitation by the river at high-water stages. A close study of other boulders in the foreground fan revealed many other cases of close fitting. This fitting explains how the positions of these boulders are so stable in the currents of flood-stage flows of the Colorado.

September 19, 1968

Kanab Point 15′ quadrangle, Arizona

September 6, 1872, morning

Grand Canyon
Camera station 876

This camera station is on the east bank at the head of Deubendorff Rapids. The view is to the north, downstream, across the head of the rapids. The location of the station was based on a dark talus block near the lower right corner of the pictures. A few other nearby blocks are recognizable, but most have been moved or rotated. Catclaw acacia is growing in the foreground and middle distance in about the same positions in both pictures. It has grown considerably on a diabase ledge in the middle distance above the center of the pictures and now obscures a more distant ledge of diabase. The water stage is twelve to fifteen feet higher in Hillers's photograph than in ours; a sand beach near the water's edge in Hillers's photograph is now much wider and is fringed by a substantial growth of tamarisk. The slope in the immediate foreground still supports a dense growth of prickly pear and barrel cactus; Mormon tea can still be seen in the lower left corner.

Note Hillers's darkroom tent and the seated figure in his picture. Members of our party are similarly dwarfed by the scale of the scene.

The low, dark canyon walls on both sides of the river are formed in the upper part of a thick, coarse-grained diabase sill intruded into the Bass Limestone. The diabase is capped by

246

light-colored ledges of dolomitic limestone and of claystone of the Bass that has been metamorphosed to hornfels. Near the center of the pictures is a slope of Hakatai Shale interrupted in the middle by a ledge of mudstone and hornfels. This slope is capped by a prominent reddish-brown cliff of Tapeats Sandstone. Above the Tapeats are a ledgy slope and overlying cliff of light-colored Bright Angel Shale. Muav Limestone forms a bench below the skyline that merges to the right with a high cliff that is mostly Redwall Limestone. Supai Group caps the Redwall in the upper right corner of the pictures and also forms the notched skyline rim in the central distance.

The talus blocks in the foreground are chiefly diabase, but many fragments of light-colored Bass dolomitic limestone are scattered among them.

In the distance beneath the notched skyline are some irregular knobs, which are on the flank of Surprise Valley. They are great blocks of Redwall Limestone in the huge landslide on which Surprise Valley was formed. The landslide is locally cemented by travertine and now forms a coherent unit, resistant to erosion. The edge of the slide can be reached from the trail to Thunder Springs; the trail begins about three miles downriver.

September 20, 1968

Powell Plateau 15' quadrangle, Arizona

September 6, 1872, morning

Grand Canyon
Camera station 883

These scenes were taken 500 feet north of the previous ones. We are on the east bank of the river at the mouth of Galloway Canyon, above the head of Deubendorff Rapids. The view is east-southeast, up Galloway Canyon. Accuracy of the camera station location is within a foot.

The stream bed in Galloway Canyon has been cut down slightly on the left, and the sand on the right has been blown or washed away. A fair-sized block, probably diabase, on the right skyline, has tumbled down. Trees in both pictures are catclaw acacia.

Rocks exposed in the lower walls of Galloway Canyon are the upper part of a coarse-grained diabase sill intruded into the Bass Limestone. Overlying beds of the Bass can be seen along the upper wall on the left and in the extreme upper right corner. Still higher beds of Hakatai Shale are framed by canyon walls in the center of the pictures. In the distance is a flank of the Powell Plateau, made up of beds ranging from the Muav Limestone to the Kaibab Formation. In the extreme left of the pictures, the end of Arrowhead Terrace (Redwall Limestone) is just visible above a Bass Limestone ledge.

248

September 20, 1968

Powell Plateau 15' quadrangle, Arizona

249

September 6, 1872, about noon

Grand Canyon
Camera station 892

This station is in Stone Creek canyon about a thousand feet from its mouth. The view is to the east-northeast up the creek and shows the waterfall. Location of the camera station, accurate to within a foot, was based on parallax between features on the canyon walls.

The foreground cliffs that support the falls are dark, coarse-grained diabase intruded into the Bass Limestone, with local granophyre. The light-colored beds above are dolomitic limestone and claystone of the Bass. Serpentine, some fibrous, is found at places in the Bass near the diabase contact. Dark, ledgy outcrops above the Bass are hornfels, mudstone, and minor sandstone beds of the Hakatai Shale. Up the canyon in the middle distance is a ledge of Tapeats Sandstone that rests unconformably on the Hakatai. Above and behind the Tapeats ledge on the right is the Redwall Limestone cliff of Arrowhead Terrace. The point on the extreme right is Redwall; higher sandstone cliffs of the Supai Group can be seen to the left of the point.

Near the center of the pictures and lower, Muav Limestone and the upper part of the Bright Angel Shale can be seen below the Redwall. Still farther in the distance, the flank of Steamboat Mountain is framed between the canyon walls (better seen in the 1872 photograph). The upper part of the

Redwall, and the Supai, Hermit Shale, Coconino Sandstone, and Toroweap and Kaibab Formations are visible on this flank.

September 20, 1968

Powell Plateau 15′ Quadrangle, Arizona

Grand Canyon
Camera Station 891

September 6(?), 1872, morning

We are still in Stone Creek canyon here, about 200 feet higher than the last camera station. We are looking west—down the canyon, across the river, and toward Great Thumb Point.

Only diabase intrusions in the Bass Limestone are shown on our side of the Colorado. Across the river, the dark, steep slope rising from the bank is also diabase. Bass above the diabase forms a light-colored bench about a third of the way up Great Thumb. The Bass is overlain unconformably by Tapeats Sandstone, which forms the dark, ragged cliffs. Bright Angel Shale makes up a series of slopes and ledges above. The Muav and Redwall Limestones form the great cliff. The peak in the center and the ledge on the right skyline are Supai Group.

September 20, 1968

Powell Plateau 15' Quadrangle, Arizona

September 7, 1872, afternoon

Grand Canyon
Camera Station 893

These views were taken from the north bank of the Colorado at the mouth of Deer Creek. We are looking to the north toward Deer Creek Falls, about 150 feet away. Our precise location of the camera station was based on foreground rocks, which are slabs of Tapeats Sandstone fallen from the cliff above. They are coated with a brown patina, apparently formed by the slightly muddy, calcareous spray from the falls. On the right side of both pictures are bushes of Mormon tea. Small clumps of fern beneath the overhanging ledge just to the right of the falls are not as numerous today, but plants are now clinging to the cliff on either side of the falls where none can be seen in Hillers's picture.

The slot from which the falls emerge appears to have been about ten feet higher in the 1872 picture. Either the flow of water was much greater then or the slot has been eroded several feet farther back into the cliff. The falls are about a hundred feet high today. No other conspicuous changes were observed in the bedrock cliff, but the pool at the foot of the falls is now shallower and much narrower. The river in 1872 may have been high enough to back up into the pool, or gravel may have built up and partly filled it. We can imagine the joy of Powell's hot, tired men when they reached this place.

The rocks in view are granite, gneiss, schist, and pegmatite, all of Proterozoic age, overlain by Tapeats Sandstone. The angular unconformity at the base of the Tapeats is about fifteen feet below the top of the falls, which is clearly visible in Stephens's picture. A mass of granite forms the buttress in front of the falls on the right. Strongly contorted gneiss and schist with inter-layered pegmatite are behind the falls.

September 22, 1968

Kanab Point 15' Quadrangle, Arizona

September 10, 1872, afternoon

Kanab Creek
Camera Station 692

This scene on Kanab Creek was the last river campsite of Powell's party. They left the boats here and rode out of the canyon on horses brought in from Kanab, Utah (as described in the introduction to this segment). Hillers took a series of photographs along the creek, of which we duplicated seven. This first one is 1,000 feet upstream from the mouth of the creek, and the view is downstream to the southeast, toward the south wall of the Inner Gorge.

Hillers's picture shows the *Emma Dean* and *Canonita*. Powell's camp no. 106 must have been close to the boats, possibly on the sand dune just above them. Our location of the camera station was based on parallax between features on the walls of Kanab Creek canyon and on the more distant canyon walls south of the river and is probably accurate to within two feet.

The bed of Kanab Creek in the foreground has changed completely. The left bank has been eroded back at least fifty feet, and none of the boulders are recognizable. The stream now flows on the extreme right side of the channel, out of sight of Stephens's camera, and the ground where the water flowed in 1872 is now bone dry. In the middle distance of our picture, a sand dune with a catclaw acacia growing at its edge has been formed by sand transported by wind from the foot of Kanab

256

Rapids. Catclaw acacia is still growing on both the right and left banks and tamarisk and willow have invaded the stream bed.

Thin ledges just above the creek are Bright Angel Shale. Tongues of Muav Limestone form the lower half of the walls seen in these pictures and the lowest cliffs across the river. Temple Butte Limestone is present above the Muav, just above the white band in the middle of the cliff. The remainder of the cliff is Redwall Limestone. A small promontory jutting from the left wall of Kanab Creek canyon near the center of the picture is capped by dissected Pleistocene talus. No conspicuous changes were noted in the canyon walls.

September 22, 1968

Kanab Point 15' Quadrangle, Arizona

September 10, 1872, afternoon

Kanab Creek

Camera Station 682

This camera station is about a quarter of a mile above the mouth of Kanab Creek. The view is northeast, upcanyon. Location of the camera station was based on parallax between features on the canyon walls and is probably accurate to within about two feet. A rock fall from a cliff behind the camera station has changed the immediate foreground completely. Blocks of sandy limestone from the lower tongue of the Muav Limestone have dropped from a niche about a hundred feet above the camera station. They apparently struck a ledge beneath the niche and scattered out across the talus slope, covering the foreground. The fall is so recent that it supports only sparse vegetation, but two young catclaw acacia trees have sprung up in the midst of the devastation. A mesquite tree is growing at about the edge of the fall near where several trees were growing in 1872 and may be a survivor from that stand. The appearance of the foreground in Hillers's photograph suggests that the site may have been subject to earlier rock falls.

Other changes have occurred in the bed of Kanab Creek. The right bank has been eroded back and steepened and the left bank may also have been steepened.

In these views, the oldest rock is a tongue of Bright Angel Shale exposed near the center of the pictures at about creek level. Elsewhere it is obscured by talus. The lowest cliff on both sides of the canyon is a tongue of Muav Limestone. A slope of Muav containing thin beds of Bright Angel forms a slight bench above this cliff. Between the bench and the small cave on the right is the Temple Butte Limestone. The upper canyon wall, above the cave, is Redwall Limestone.

September 22, 1968

Kanab Point 15' Quadrangle, Arizona

Mid-September 1872, early afternoon

Kanab Creek
Camera Station 685

We have now come up Kanab Creek about three-and-a-quarter miles from its mouth, and our view is to the northeast, upstream. Location of the camera station was based on foreground blocks. The different time of day of our picture resulted in unavoidable deep shadow in the foreground, and the description that follows is based on our observations at the scene rather than on features discernible in the photograph.

All the largest rocks are still in place, but the smaller blocks and boulders are gone. At least three blocks of Muav Limestone in the center of the pictures are more than fifteen feet across. A cherty limestone block just in front of Hillers's camera and another large one on the right are Redwall Limestone. A dark limestone block of Muav or Temple Butte Limestone lies near the lower center of the pictures about ten feet from the camera station; its upper corner has been chipped off since 1872. Another block of Muav in the lower left corner of Hillers's picture is now gone and the bank has been washed out and steepened somewhat on that side.

A green shaly limestone tongue of the Muav forms a cliff and overlying slope just above the water on the left, behind large blocks. Above it lies a stepped cliff of either the upper tongue

260

of the Muav or the Temple Butte. Red-wall forms the large cliff in the center of the pictures. No conspicuous changes are evident in the canyon walls.

September 23, 1968

Kanab Point 15' Quadrangle, Arizona

Mid-September 1872, about noon

Kanab Creek
Camera Station 638

We are about four miles above the mouth of Kanab Creek, looking northwest, upcanyon. Location of the camera station was based on parallax between features observed on the canyon walls, and it is probably accurate to within two feet. The bank on the left has been cut back and steepened, and the character of the stream bed has changed completely. The large blocks in the stream bed are gone and a new crop of boulders has been deposited.

The oldest rock visible is Muav Limestone, which forms the low ledges up the canyon and across the stream. The banded cliffs on both sides of the canyon are Temple Butte Limestone. The high, sheer cliff of Redwall Limestone is framed between the Temple Butte cliffs. We noticed no conspicuous changes in the canyon walls.

262

September 23, 1968

Kanab Point 15' Quadrangle, Arizona

Mid-September 1872, afternoon

Kanab Creek
Camera Station 578

This camera station is about four-and-a-half miles above the mouth of Kanab Creek, and we are looking south-southeast, down the canyon. Location of the camera station was based on parallax of features on or near the canyon walls, but Stephens's station was offset forward and to the left about two feet to avoid a large clump of prickly-pear cactus. Both stations were on a sandy alluvial terrace on the east bank of the creek. The terrace provides an ideal environment for prickly pear, as well as for catclaw acacia, grasses, and Mormon tea shrubs. The scene has hardly changed since 1872.

The lower ledgy cliff on the left canyon wall is Temple Butte Limestone. Remnants of two terrace deposits of probable Pleistocene age rest on the cliff but are not obvious in the pictures. Higher on both canyon walls and in the center is the great cliff of Redwall Limestone capped by lowermost Supai Group. No conspicuous changes were noted in the canyon walls.

September 23, 1968

Kanab Point 15' Quadrangle, Arizona

September 1872, early afternoon

Kanab Creek
Camera Station 647

Still higher on Kanab Creek, we are here about six miles above its mouth. The view is to the northeast, up the canyon. Location of the camera station was based on parallax between features on talus in the middle distance and on the canyon walls, and also on the foreground slab of cherty Redwall Limestone that was partly obscured by vegetation in 1872. Many of the great blocks of limestone on the canyon floor have been rotated, but the largest, slightly to right of center, is still in the same position. One of the rotated blocks is about twenty feet across its diagonal. The source of a large slab that has fallen shows clearly as a rectangular outline in the sunlit cliff in the upper left corner of Stephens's picture. New rock fragments in the middle distance were derived from this slab.

The lower part of the Redwall Limestone forms the visible part of the wall on the left. Redwall also forms the great cliff, whose cap of Supai Group is just visible through a notch, left of center, in the far canyon wall.

September 24, 1968

Kanab Point 15' Quadrangle, Arizona

September 1872

Kanab Creek
Camera Station 71-b

This last camera station on Kanab Creek is about seven miles above its mouth; the view is downstream to the west-southwest. We located the camera station on the basis of parallax of features on the pinnacle and more distant features of the canyon walls. The positions of the boulders, except for two very large ones downstream, have changed. Catclaw acacia is growing high on each bank of the stream today as in Hillers's picture, but bear grass is much less abundant today.

The ledges low on each canyon wall and the cliffs at the base of the pinnacle are Temple Butte Limestone. The pinnacle itself is Redwall Limestone, as is most of the great cliff behind it. Above the cliff are ledges of lower Supai Group. No conspicuous changes were noted on the canyon walls.

September 24, 1968

Kanab Point 15' Quadrangle, Arizona

April 19 or 20, 1872

Grand Canyon

Camera Station 597

We are now back on the Colorado River, on the northwest bank about 800 feet above Lava Falls. The view is downriver to the southwest, toward the falls. Fennemore and Hillers took this picture and the two that follow on an overland trip from the North Rim to Lava Falls earlier in the year. Exact location of the camera station was based on foreground rocks and a bed-rock ledge shown at the bottom of the pictures, but the water is one to two feet higher in our photograph. Many of the rocks in the middle distance on the right are obscured by tamarisk, and catclaw acacia grows thickly above these rocks today as it did in 1872.

The rocks in view range from Tapeats Sandstone to Pleistocene basalt and gravel. The uppermost part of the Tapeats forms the foreground ledge. Downstream on the left, Muav Limestone forms most of the bold cliff extending up from the river. Between the camera station and this cliff, the beds are downdropped to the west across the Toroweap fault. Temple Butte Limestone forms ledges just below the narrow light-colored bench high on the wall. Redwall Limestone forms the uppermost part of this cliff. A patch of basalt is perched on a bench on the Muav on the left side of the photographs. The slope extending to the upper left corner of the pictures is shale

270

and limestone of the Supai Group. The right canyon wall is basalt with local interlayered gravel near the base, and some Muav is exposed between two main remnants of lava flows. In the far distance is a dark mass of basalt that flowed down over the edge of the Inner Gorge. The edge of a dissected Pleistocene boulder fan can be seen just above the river on the left.

The basalts that poured over the rim of the canyon here are rich in peridotite nodules, which can be found in many of the boulders along the river bank, including those in the foreground. Some of these basalts are very fine grained to glassy and may have been chilled by river water as they solidified.

September 26, 1968

Vulcans Throne 7½' Quadrangle, Arizona

April 19 or 20, 1872

Grand Canyon
Camera Station 693

This camera station is also on the northwest bank of the river at Lava Falls. The downstream view is to the southwest, across the lower part of the rapids. Location of the camera station was based on large foreground boulders of basalt whose surfaces are intricately sculptured. The largest boulders, except for one on the right, are still in place. This missing boulder was the one from which Hillers took the next picture. The sculptured basalt blocks are very fine grained to glassy and carry inclusions of peridotite, granite, and sedimentary rocks. On the far bank, springs emerge from the base of a Pleistocene boulder fan. Vegetation is luxuriant at the springs; the bright spots near the springs are where the sunlight was reflected from a wet mossy bank of travertine.

The water stages in the two pictures are almost identical. Many of the same holes and other features of the rapids can be recognized.

September 26, 1968

Vulcans Throne 7½' Quadrangle, Arizona

April 19 or 20, 1872

Grand Canyon
Camera Station 515

At our last camera station, we are still on the northwest bank, near the middle of Lava Falls. The view is upstream to the northeast, showing the head of the falls. The great block of basalt on which Hillers's camera station was located has been washed away, but we occupied another block twenty to thirty feet farther back that permits a similar perspective.

Bright Angel Shale is exposed in the lower part of the right canyon wall and forms a small pinnacle to the right of center (in shadow in Hillers's picture). Most of the Bright Angel is covered with Pleistocene talus and landslide debris. The remainder of the right canyon wall (in shadow in Hillers's photograph) is Muav and Redwall Limestones. On the butte in the center framed by the canyon walls, Muav and Temple Butte Limestones form the lowest visible stepped cliff. The smooth cliff above it is Redwall. The Supai Group forms the overlying slope and the cliffs extending to the skyline. The left canyon wall is basalt, as are the great blocks that form the falls.

274

September 26, 1968

Vulcans Throne 7½' Quadrangle, Arizona

Table VII: Geologic Periods and their Approximate Duration

The completeness of the geologic record in the canyons traversed by the Powell expedition of 1871-72 can be seen below. Rocks of Archean age and older—the most ancient rocks known worldwide—have not been found in the Grand Canyon, and nondeposition or erosion in the region is responsible for the absence of Ordovician and Silurian rocks. All other periods, however, are represented by at least one geologic formation that is noted in the figure captions.

Eon	Era	Period		Epoch	Age estimates of boundaries (million years)
Phanerozoic	Cenozoic	Quaternary		Holocene / Pleistocene	2
		Tertiary	Neogene	Pliocene / Miocene	24
			Paleogene	Oligocene / Eocene / Paleocene	63
	Mesozoic	Cretaceous			138
		Jurassic			205
		Triassic			240
	Paleozoic	Permian			290
		Carboniferous	Pennsylvanian		330
			Mississippian		360
		Devonian			410
		Silurian			435
		Ordovician			500
		Cambrian			570
Proterozoic	Late				900
	Middle				1,600
	Early				2,500
Archean	Late				3,000
	Middle				3,400
	Early				(3,800?)
pre-Archean					4,550

Glossary

Alluvium   Gravel, sand, silt, or clay deposited by a stream. An **alluvial fan** is a fan-shaped accumulation of sediment at the mouth of a canyon.

Anticline   A convex-upward fold in rocks. (A **syncline** is a convex-downward fold.)

Arkose   Feldspar-rich sandstone, commonly coarse grained.

Basalt   Dark, very fine to fine-grained volcanic rock composed chiefly of iron-magnesium minerals; most common as lava flows.

Bedrock   Solid rock that underlies soil or other unconsolidated material.

Biotite   Black or near-black, iron-magnesium mineral of the mica group; common as flaky grains in granitic rocks.

Breccia   Rock composed of angular rock fragments cemented together; differs from **conglomerate** in that edges of fragments are sharp.

Calcareous   Limy; contains calcium carbonate.

Chert   Hard, dense sedimentary rock composed of microscopic quartz crystals; light or dark colored in various hues; commonly called **flint** if used for artifacts such as arrowheads.

Colluvium   Loose slope wash of soil and rock fragments.

Conglomerate   Sedimentary rock; consolidated gravel composed of water-rounded rock fragments ranging in size from granules to boulders; fragments are cemented by fine-grained minerals or clay.

Crossbedding   In sedimentary rock, an original arrangement of beds at an angle to the horizontal layering. Characteristic of Coconino Sandstone, in which it is the result of migration of sand dunes before they became rock.

Desert Varnish   Dark, bluish- or brownish-black, shiny coating, mainly of iron or manganese oxides, on rock surfaces in desert areas.

Diabase   Fine-grained, dark-colored igneous rock intruded at shallow depth.

Diorite   Fine- to coarse-grained igneous rock, commonly gray; formed at depth.

Dike   Tabular intrusion of igneous rock that crosscuts layering of preexisting rock.

Dip   Angle made by a fault or a bed of rock with a horizontal plane.

Disconformity   Unconformity at which overlying and underlying beds are parallel; generally represents a considerable gap in time marked by erosion or nondeposition.

Dripstone   Calcium carbonate deposited by dripping water; stalactites and stalagmites are common forms.

Fault   Fracture in rock along which movement has occurred. May result in downdropping of a younger unit to the level of an older one (as at camera station 680 in segment 2), in upthrusting of an older rock to the level of a younger, or in lateral displacement.

Feldspar   Common rock-forming mineral, a sodium-calcium or potassium aluminum silicate; light colored.

Foliation   Bending or layering in metamorphic rocks.

Gneiss   Coarse-grained metamorphic rock characterized by alternating bands of light- and dark-colored minerals.

Granite   Coarse-grained, crystalline, igneous rock formed at great depth; nonbanded; quartz and feldspar are dominant materials.

Granophyre   Granitic rock characterized by some large crystals surrounded by very small crystals.

Gypsum   Light-colored soft mineral, a hydrous calcium sulfate. Like rock salt, with which it is commonly associated, it is formed by evaporation.

Hornfels   Very fine grained metamorphic rock; nonbanded; commonly formed around igneous intrusions.

Igneous Rock   Solidified from magma (molten rock). Includes the lavas of the Inner Gorge. **Igneous** rocks form one of the three main groups into which rocks are divided. (The other two are **sedimentary** and **metamorphic**.) The term is also applied to minerals.

Limestone   Sedimentary rock that is more than 50% calcium carbonate.

Marlstone   Clayey limestone.

Meander   Bend in a mature stream produced by its swings from side to side in a flood plain.

Metamorphic Rock   Sedimentary or igneous rock that has been metamorphosed (changed) by heat and/or pressure, usually at great depth.

Mudstone   Solidified mixture of clay and silt; similar to shale but does not break into thin layers.

Parallax   Apparent displacement of an object as seen from two points not on a straight line with the object.

Pegmatite   Igneous rock containing very coarse grains and interlocking crystals; composition is generally granitic; forms irregular lenses, veins, or intrusions that cut across bedding of surrounding rock.

Quartz   Crystalline silica (SiO_2); generally colorless, white, or colored by impurities. Common rock-forming mineral; the major constituent of sand and sandstone.

Quartzite   Very hard, homogeneous sandstone composed of quartz grains firmly cemented by silica.

Sandstone   Sedimentary rock composed of sand-size grains, usually of quartz; deposited by water or wind.

Schist   Metamorphic crystalline rock characterized by closely spaced bands; splits easily; mineral composition is varied.

Sedimentary Rock   Formed by solidification of loose sediments deposited in layers. Sedimentary rocks mentioned in text include limestone, conglomerate, and sandstone.

Serpentine   Hydrous magnesian silicate found in igneous and metamorphic rocks; generally in shades of green; greasy or silky luster; soapy feel; soft enough to cut with knife.

Shale   Sedimentary rock formed by consolidation of mud, clay, or silt; characterized by laminations (thin beds) along which it splits easily.

Sill   Tabular igneous intrusion that roughly parallels the layers of enclosing rock.

Silt, Siltstone   **Silt** is composed of generally unconsolidated rock fragments intermediate in size between very fine sand grains and coarse clay. **Siltstone** is a sedimentary rock that commonly contains some clay and fine sand; it is similar to shale but is not laminated.

Spall   Fragment of rock broken off by erosion or hammering. The word is also used as a verb.

Talus   Accumulation of angular rock fragments at the base of a cliff or steep slope from which they were derived.

Terrace   Roughly planar surface, commonly long and narrow, bordered by steeper slopes. A terrace is commonly erosional (cut by waves) or depositional (formed by water-laid sediments).

Travertine   A form of limestone deposited from solution; generally white or tan. When deposited slowly, to form stalactites and stalagmites in caves, it is hard, dense, and finely crystalline. When deposited quickly, as by springs, it is softer, spongy (**tufa**).

Tufa   The softer, spongy form of **travertine** (see above), commonly formed by springs.

Unconformity   Gap in the rock record due to a considerable period of nondeposition or erosion; the rock surface that represents this gap. For example, Table VI shows a major unconformity between the Cambrian formations and the Devonian Temple Butte Limestone, a gap of about 100 million years. At an **angular unconformity**, younger flat-lying rocks have been deposited over older beds that have been tilted and then eroded; an example is the unconformity between the older Nankoweap Formation and the younger Tapeats Sandstone at camera station 451, segment 5.

Selected Reading

General Interest

Baker, Pearl, 1970, Trail on the water: Boulder, Colo., Pruett Press, 134 p.

Butchart, Harvey, 1970, Grand Canyon treks—a guide to the inner canyon routes: Glendale, Calif., La Siesta Press, 72 p.

————1975, Grand Canyon treks II—a guide to the extended canyon routes: Glendale, Calif., La Siesta Press, 48 p.

Darrah, W.C., 1951, Powell of the Colorado: Princeton, Princeton University Press, 426 p.

Dellenbaugh, F.S., 1902, The romance of the Colorado River: 1965 reprint, Chicago, Rio Grande Press, 399 p.

————1908, A canyon voyage: New York, Putnam; 1926 reprint, New Haven, Yale University Press; 1984 reprint, Tucson, University of Arizona Press, 277 p.

Fowler, D.D., ed., 1972, Photographed all the best scenery, Jack Hillers' diary of the Powell expeditions, 1871-1875: Salt Lake City, University of Utah Press, 225 p.

Fowler, D.D., Euler, R.C., and Fowler, C.S., 1969, John Wesley Powell and the anthropology of the canyon country: U.S. Geological Survey Professional Paper 670, 30 p.

Fradkin, P.L., 1981, A river no more—the Colorado River and the West: New York, Knopf, 360 p.

Freeman, L.R., 1923, The Colorado River, yesterday, today, and tomorrow: New York, Dodd, Mead, 451 p.

Freeman, L.R., 1924, Down the Grand Canyon: New York, Dodd, Mead, 6 p.

Hughes, J.D., 1978, In the house of stone and light: Grand Canyon, Grand Canyon Natural History Association, 137 p.

Kolb, Ellsworth, 1947, Through the Grand Canyon from Wyoming to Mexico: New York, Macmillan, 344 p.

Lavender, David, 1985, River runners of the Grand Canyon: Grand Canyon, Grand Canyon Natural History Association, 147 p.

Powell, J.W., 1875, Exploration of the Colorado River of the West and its tributaries: Washington, U.S. Government Printing Office, 291 p.

————1895, Canyons of the Colorado: Meadville, Pa., Flood and Vincent [reprinted in 1961 by Dover Publications, New York, under the title "The exploration of the Colorado River and its canyons"], 400 p.

Powell Society, Ltd., publications: River runners' guides to the Green and Colorado Rivers, with emphasis on geologic features.
1. Hayes, P.T., and Simmons, G.C., 1973, Dinosaur National Monument and vicinity, 78 p. This volume replaces both the original Volume I (Hayes, P.T., and Santos, E.C., 1969, From Flaming Gorge Dam through Dinosaur Canyon to Ouray), now out of print, and Hayes, P.T., 1971, Yampa River supplement to Volume I.
2. Mutschler, F.E., 1977, Canyonlands National Park and vicinity, 99 p. This volume replaces the original Volume II (Mutschler, F.E., 1969, Labyrinth, Stillwater, and Cataract Canyons), now out of print.
3. Simmons, G.C., and Gaskill, D.L., 1969, Volume III, Marble Gorge and Grand Canyon, 132 p.
4. Mutschler, F.E., 1972, Volume IV, Desolation and Gray Canyons, 85 p.

Shoemaker, E.M., and Stephens, H.G., 1975. First photographs of the canyon lands, in Fassett, J.E., ed., Canyonlands country, A guidebook of the Four Corners Geological Society, p. 111-122.

Stanton, R.B., 1932, Colorado River controversies. New York, Dodd, Mead, 232 p.

Stegner, Wallace, 1954, Beyond the hundredth meridian—John Wesley Powell and the second opening of the West: Boston, Houghton Mifflin, 438 p.

Turner, R.M., and Karpiscak, M.M., 1980, Recent vegetation changes along the Colorado River between Glen Canyon Dam and Lake Mead, Arizona: U.S. Geological Survey Professional Paper 1132, 125 p.

U.S. Geological Survey, 1969, The Colorado River region and John Wesley Powell: U.S. Geological Survey Professional Paper 669, 145 p.

Utah State Historical Society, 1947, The exploration of the Colorado River in 1869: Utah Historical Quarterly, v. 15, 270 p.

————1948 and 1949, The exploration of the Colorado River and the high plateaus of Utah in 1871-72: Utah Historical Quarterly, v. 16 and 17, 540 p.

Zwinger, Ann, 1975, Run, river, run; a naturalist's journey down one of the great rivers [the Green] of the American West: New York, Harper and Row, 317 p.

Geology

Breed, J.W., and Roat, E.C., eds., 1974, Geology of the Grand Canyon: Flagstaff, Museum of Northern Arizona, 185 p.

Culbertson, W.C., 1962, Laney Shale Member and Tower Sandstone Lentil of the Green River Formation, Green River area, Wyoming: U.S. Geological Survey Professional Paper 450-C, p. C54-C57.

Dutton, C.E., 1882, Tertiary history of the Grand Canyon district: U.S. Geological Survey Monograph, v. 2, 264 p.

Gregory, H.E., 1938, The San Juan country: A geographic and geologic reconnaissance of southeastern Utah: U.S. Geological Survey Professional Paper 188, 123 p.

Hansen, W.R., 1965, Geology of the Flaming Gorge area, Utah-Colorado-Wyoming: U.S. Geological Survey Professional Paper 490, 196 p.

————1969, The geologic story of the Uinta Mountains: U.S. Geological Survey Bulletin 1291, 144 p.

Hansen, W.R., Rowley, P.D., and Carrara, P.E., 1983, Geologic map of Dinosaur National Monument and vicinity, Utah and Colorado: U.S. Geological Survey Miscellaneous Investigations Series Map I-1407, scale 1:50,000.

Huntoon, P.W., Billingsley, G.H., Breed, W.J., 1982, Geologic map of Canyonlands National Park and vicinity, Utah: Moab, Utah, The Canyonlands Natural History Association, scale 1:62,500 [2 sheets].

Huntoon, P.W., Billingsley, G.H., Breed, W.J., Clark, M.D., and others, 1976, Geologic map of the Grand Canyon National Park, Arizona: Grand Canyon, The Grand Canyon Natural History Association, scale 1:62,500.

Kieffer, S.W., 1987, The rapids and waves of the Colorado River, Grand Canyon, Arizona: U.S. Geological Survey Open-File Report 87-096.

Lohman, S.W., 1974, The geologic story of Canyonlands National Park: U.S. Geological Survey Bulletin 1327, 126 p.

McKee, E.D., 1982, The Supai Group of Grand Canyon: U.S. Geological Survey Professional Paper 1173, 504 p.

Peterson, Fred, and Pipiringos, G.N., 1979, Stratigraphic relations of the Navajo Sandstone to Middle Jurassic formations, southern Utah and northern Arizona: U.S. Geological Survey Professional Paper 1035-B, 43 p.

Pipiringos, G.N., and O'Sullivan, R.B., 1978, Principal unconformities in Triassic and Jurassic rocks, western interior United States—a preliminary survey: U.S. Geological Survey Professional Paper 1035-A, p. A1-A29.

Poole, F.G., and Stewart, J.H., 1964, Chinle Formation and Glen Canyon Sandstone in northeastern Utah and northwestern Colorado: U.S. Geological Survey Professional Paper 501-D, p. D30-D39.

Camera Station Index

Index

Hal G. Stephens

Hal G. Stephens was born in 1917 in a town formerly known as Chefoo in northern China. His father was employed in the marine shipping business, and both of his paternal grandparents were missionaries in China.

After earning an M.A. in geology from the University of Missouri and brief employment with Houston-based oil companies, he joined the United States Geological Survey in 1942, where he remained until his retirement in 1977. Following service with the Army Air Corps, Stephens taught at the University of Missouri. His early work with the USGS involved investigations and mapping of various mineral deposits around the country.

Throughout his career, Stephens maintained an absorbing interest in photography. About 1960 he and E.C. Morris made copies of the best lunar photographic plates taken by the Mount Wilson observatory in the late 1920s. These plates provided the basis for the earliest lunar geologic mapping initiated by E.M. Shoemaker of the USGS. In 1964 he transferred to the lunar and planetary program of the Survey in Flagstaff, Arizona, where he organized a photographic unit to document astronaut training exercises in preparation for NASA's manned flight to the moon in 1969. He later developed and operated the Lunar and Planetary Distribution Facility, which contained thousands of photographs obtained from NASA's lunar and planetary missions. His films of the effects of the great Alaska earthquake of 1964 were the basis for the USGS film, "The Good Friday Earthquake of 1964."

The river expedition upon which this book is based initiated his river-running experience. Many other exciting trips have followed. Following Stephens's retirement from the USGS in 1977, the life and work of John Wesley Powell has been a major interest, and he has devoted much of his time to the preparation of this album of comparative photographs. He lives in Flagstaff, Arizona.

Eugene M. Shoemaker

Eugene M. Shoemaker was born in Los Angeles in 1928. He attended the California Institute of Technology and Princeton University, where he received his Ph.D. in 1960.

A geologist with the United States Geological Survey since 1948, he was asked to form and head the Survey's Branch of Astrogeology in 1961. He has made the study of moons his specialty. In the early 1960s he was a co-investigator for the Ranger lunar probe series and acted as principal investigator, geological field investigations, in the Apollo lunar landings, 1965-1970. Since 1978 he has been a co-investigator for NASA's Voyager missions to Jupiter and Saturn.

Dr. Shoemaker has taught at the California Institute Technology and from 1969 to 1972 was chairman of the Division of Geological and Planetary Sciences at that institution. His current research is on impact processes in the solar system and the effects of large body impacts on the evolution of life.

Closer to home, Dr. Shoemaker shares an interest with the other members of the Powell Society of the life and work in the great pioneering geologist and explorer, John Wesley Powell. When his work does not take him away, he lives near Flagstaff, Arizona.